Basic
Home Wiring
Illustrated

By the Editors of Sunset Books and Sunset Magazine

Lane Publishing Co. • Menlo Park, California

Acknowledgments

It is with sincere appreciation that we thank the following people for contributing their special talents toward the creation of this book: Fred Barson, John McClements, John B. Sallemi, Edward Selden, Diane Tapscott, and Donald W. Vandervort.

For sharing their knowledge and expertise, we also extend thanks to John E. Denkhaus, Sr., Journeyman Electrician; Donald E. Johnson, Building Inspector, City of Menlo Park; Paul Y. Lin, Director Technology Division, College of San Mateo; Louis Electrical Supply; Arel Sessions, Electrical Safety Inspector, State of Idaho; and Stanford Electric Works.

Edited by Linda J. Selden

Technical Consultants: William Bailey, Senior Electrical Inspector, City of Santa Clara, California

Eugene Brown, Senior Electrical Inspector, City of Palo Alto, California

T. R. Parkhill

Design and Illustrations: Ted Martine

Cover: Photographed by Norman A. Plate

Editor, Sunset Books: David E. Clark

Fourth Printing January 1980

Contents

Special features

An introduction to the basics

Keeping our homes warm and our ice cream cold, electricity provides us generously with comfort and conveniences. And fortunately for you as an amateur, electrical work is one of the easiest kinds of home maintenance and repair. It is simple and neat. It doesn't require a shop full of specialized tools. And there is considerable standardization in home electrical systems and related materials. But before you embark on any electrical wiring projects, it's important to understand a few things about electricity itself, electrical safety, and electrical codes. This first chapter gives you this general background.

The current course

Visualize a stream. As youngsters we watched the flowing water carry leaves and twigs. In a similar way, you can think of electricity as a current of very tiny particles (electrons) flowing through a wire. Associated with this flowing current are three terms you will work with often: *watts, volts,* and *amperes.*

DEFINING THE TERMS

Electric power, the energy per second supplied by the current (to light a light bulb, for instance), is expressed in *watts.* The potential difference, or pressure, causing the current to flow is measured in *volts.* The amount of current that flows through the wire or device is measured in *amperes* (amps). For the purposes of this book, the relationship among these units is represented in the simple formula *volts × amperes = watts.*

A glossary defining many other terms you might need to know is provided on pages 86-87.

PATH OF LEAST RESISTANCE

Another important characteristic of electric current is that it is choosy about the materials it flows through. It is partial to flowing in the path offering the least resistance.

The general term "conductor" applies to anything that permits, or conducts, the flow of electricity rather than resisting it. Certain materials make better conductors than others. Copper, for example, is a good conductor. Rubber, on the other hand, is a very poor conductor, offering so much resistance that it's often used as an insulator to prevent any flow of electricity between conductors.

The circuit loop

Now that we're thinking of electricity as a current flowing through conductors, let's look at a second basic concept: the continuous loop of a circuit.

In order to flow, electric current must have a continuous path from start to finish — like a circle. The word "circuit" refers to the entire course an

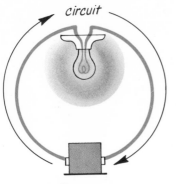

circuit

voltage source

electric current travels, from the source of power through some device using electricity (such as a toaster) and back to its starting point, the source. So what may appear to be a hopelessly tangled maze of wires running through the walls and ceilings of your home is actually a well-organized system composed of several circuits.

Each circuit forms a continuous closed path that can be traced from its beginning in the service entrance panel or subpanel through various outlets or appliances and back to the service entrance panel or subpanel.

Codes and permits

Before you get too involved with your wiring project — whether you're updating the wiring in an older home, wiring a new home, or adding a 120/240 volt circuit for a newly purchased clothes dryer — there are two local organizations you should keep in mind: your building department and your utility company. Both have important roles in any electrical work you do beyond minor repair or replacement of fixtures.

As a homeowner interested in doing your own wiring, your first step should be to check with your municipal building department. There the electrical inspector will talk with you about local codes, the National Electrical Code, and your jurisdiction's requirements concerning permits and inspections.

The National Electrical Code (referred to as "NEC" or simply "the Code") is a set of rules spelling out wiring methods and materials to be used in electrical work. With safety as its purpose, the NEC forms the basis for all regulations applied to electrical installations.

Some cities, counties, and states amend the NEC to suit their particular purposes. As a result, specific regulations can differ from county to county and even from town to town. If you move to a new location, don't expect the electrical code there to be identical to the one you left behind — some variations are almost sure to exist.

The codes establish the standards you must meet. *Basic Home Wiring Illustrated* gives you the practical information you need to work within the framework of the NEC.

Canadian requirements. In Canada, some requirements differ from those listed in this book, in particular those for grounding, service entrance, and acceptable circuit loads. Check the Canadian Electrical Code.

THE UTILITY COMPANY

Your wiring plans may involve a change in electrical service — for example, your home may have 120-volt, two-wire service and you need 240-volt, three-wire service for a new clothes dryer. If so, you must contact your local utility company to obtain the additional service.

You should also contact your utility company whenever your building or remodeling plans call for changing the location of your meter and service entrance panel. The company will advise you on proper placement of service entrance equipment for best connections to supply lines.

Even if you already have 240 volts and the location of service suits you, checking with the utility company in advance is a good idea if your project will increase your load. You may avoid trouble later by making sure that your utility company cables are heavy enough for your new load.

AMATEUR VS. PROFESSIONAL ELECTRICIAN

Doing your own electrical work may not always be the best idea. A check with the building department may reveal that your jurisdiction has restrictions on how much and what kinds of electrical wiring a homeowner may do. For instance, you may be able to do all wiring up to the point at which the circuits are connected to the service panel, but the final hookup may have to be done by a licensed electrician.

Even if your locality does not restrict what you may do in your own home, you may wish to use the services of a professional electrician. If things crop up that you don't understand or if there's some doubt remaining in your mind, it's best to call on a professional.

Working safely with electricity

Electricity — something we all use freely with just a flick of a switch. Electricity — something to be treated with caution and respect.

While these two statements seem to express conflicting viewpoints, together they give a good foundation for working safely with electricity. Once you understand and respect the potential hazards, electrical wiring is quite safe to do.

POTENTIAL HAZARDS

Most of us have heard stories of fires or injuries from causes related to electricity. Barns burn to the ground because of electrical storms, homes are destroyed because of faulty wiring, people are shocked — or in extreme cases, killed — by household electrical accidents. We have no control over electrical storms, but we can and must be careful with electricity in our homes.

Fires. Several faulty wiring situations can cause fires. For example, restriction of the flow of current through a wire or cord, as when a cord is poorly connected to its plug, may lead to overheating and eventually to a fire.

Another fire hazard frequently found in homes is the "extension cord octopus." This results when too many appliances are plugged into an extension cord that isn't hefty enough to carry the electricity these appliances demand. Excessive heat builds up in the cord as it tries to carry the load for all the appliances; the cord's insulation becomes brittle or melts from the heat; wires are exposed as the insulation deteriorates; and eventually a short circuit develops, sending sparks flying when the bare wires touch each other.

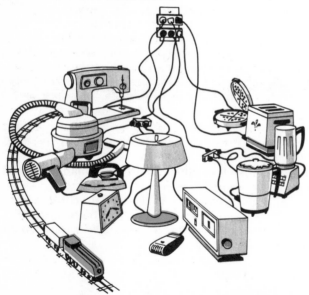

Too much demand on receptacles and/or extension cords creates a fire hazard.

Shocks. Does the thought of electric shock trigger a twinge of fear in you? If it does, you have lots of company. Let's make sure, though, that this fear isn't based on misconceptions — like the fear of

getting warts from touching a toad. Only under certain conditions can you get an electric shock; just touching a wire won't necessarily hurt you.

You'll recall from the discussion of circuits on pages 4 and 5 that current flows in a continuous, closed path, from its source through a device that uses the power and then back to the source. If you accidentally become a link in an electrically live circuit, you'll get a shock.

The key word here is "link." To get shocked, you must be touching a live wire or device. At the same time, you must also be touching a grounded object or another live wire.

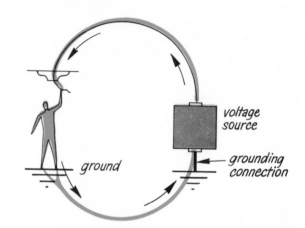

Current passes through man when he becomes a link in an electrically live circuit.

It is important to realize that electricity need not flow in wires to make the return trip to its source. It can return to the source through any conducting body — including you — that contacts the earth directly or touches a conductor that in turn enters the earth.

This may sound like a rather unlikely situation. But consider that whenever you're partially immersed in water, touching any metal plumbing fixture, or standing on the ground or on a damp concrete basement, garage, or patio floor, you're in contact with a grounded object. In other words, you're satisfying one of the two requirements for getting a shock.

There may be two requirements for getting a shock, but there's only one requirement for not getting a shock: make sure that the circuit you intend to work on is dead.

KILL THE CIRCUIT FIRST

The most important rule for all do-it-yourself electricians is this: never work on any electrically "live" circuit, fixture, or appliance. Your life may depend on following this rule.

Before starting any work, you must disconnect (or "kill") the circuit you'll be working on at its source in the service panel. If your circuits are protected by fuses, removing the appropriate fuse disconnects the circuit from incoming service. In a service panel equipped with circuit breakers, you can disconnect a circuit by switching its breaker to the OFF position.

For more information on working with fuses and circuit breakers, see pages 22–23.

To make sure you disconnect the correct circuit, turn on a light that's connected to the circuit before you remove the fuse or turn off the circuit breaker. The light will go out when you've removed the correct fuse or turned off the correct breaker.

If you have any doubt about which fuse or breaker protects which circuit, shut off all current coming into your home at the main disconnect (see "Shutting off main power supply," page 10).

While you're at your service panel, spend another moment to prevent a possible disaster. Tape a note on the panel explaining what you're doing so no one will come along and replace the fuse or reset the circuit breaker while you're working on the wiring. Then either carry the fuse with you in your pocket or tape the circuit breaker in its OFF position.

Leave note on service panel to alert others when you are working on circuit wiring.

One final step before starting your work is to check that the circuit is actually dead. Using a night light or a test lamp (see page 36), test the circuit. If the light doesn't glow when switched on, the circuit is dead. But if there is light, return to your service panel; you've killed the wrong circuit.

When your work involves only a portable electric appliance, unplug it first. Just turning off the power switch is *not* enough.

Put down dry boards to stand on when doing any electrical work in damp locations.

MORE SAFETY PRECAUTIONS

With the electricity turned off, you can work in complete safety. Still, it's a good idea to keep a few additional safety precautions in mind.

Remember that water and electricity don't mix. Never work on wiring, fixtures, switches, or appliances when you're wet or standing on a damp spot. Lay down dry boards to stand on if the floor or ground is wet.

Be sure you thoroughly understand how your particular home is wired before you modify or work on the electrical system. The procedures described in this book are based on the assumption that your existing wiring was done correctly.

Remember that your circuits are dead only past the points where they have been opened or disconnected. In particular, the lines from the utility company in your service entrance panel are still hot, even after the fuses are removed or the circuit breakers turned off.

Throughout the book we'll be reminding you of various safety precautions. Each one will be presented in the following manner:

△ Make sure circuit you're going to work on is dead. Test it before making any repairs or connections.

How electricity energizes your home

To gain a better understanding of the way a home electrical system works before you get into wiring, read on as we untangle the maze of wires, starting at the point where the utility company supplies your home with electricity.

From transformer to toaster

Until a fuse blows or a storm leaves us eating a cold dinner by candlelight, many of us take electricity — and the conveniences it powers — for granted. This is natural, because electricity seems to come so easily. We just telephone the local utility company to have electricity brought to our homes.

YOUR ELECTRICAL SERVICE

Utility companies distribute power to individual households through overhead wires or underground cables. Today, most homes have three-wire service. That is, the power company connects three lines to your service entrance equipment. With this arrangement, there are two "hot" conductors (wires), each supplying electricity at 120 volts with respect to one "neutral" conductor. During normal operation, this neutral wire is maintained at zero volts, or what we refer to as "ground potential."

Three-wire service provides both 120-volt and 240-volt capabilities. One hot conductor and the neutral combine to provide for 120-volt needs, such as lights or wall receptacles (outlets). Both hot conductors combine with the neutral to provide 120/240 volts for such large appliances as a range or clothes dryer.

Many older homes have only two-wire service. These homes have only one hot conductor at 120 volts and a neutral conductor.

An explanation is in order about the designation of the voltage supplied by the power company. As we mentioned in the previous chapter, voltage is electrical pressure. Furthermore, this pressure can fluctuate from roughly 115 volts to 125 volts, even within the same day. That is why you may see references elsewhere to household voltages other than 120. Throughout this book we will use 120 as the voltage for each hot line supplied by the power company.

How do three wires energize your home?

Now let's see what actually happens in an electrical system and why it happens. In a step-by-step way, we will trace the electrical path through your home, starting at the meter.

Meter. As shown in drawing **9-A**, electricity passes through a meter before it enters your service entrance panel.

Owned, installed, and serviced by the utility company, a meter is the final step when you put in a complete wiring system. Once installed, the meter measures the electrical energy your household consumes in kilowatt-hours. ("Kilowatt-hours" refers to the rate of energy consumption in kilowatts multiplied by the time of usage in hours.)

To read and interpret your meter, see "How to read your meter," page 9.

Service entrance panel. The control center for your electrical service is the service entrance panel. In this panel — a cabinet or box — you'll usually find the main disconnect (the main fuses or main circuit breaker), the fuses or circuit breakers protecting each individual circuit in your home, and the grounding connection for your entire system.

Looking at drawing **9-A** again, follow the three conductors from their connection with the power company lines, through the meter, and into the service entrance panel. Once inside the service panel, the two hot conductors (red, black, or any other color except white, gray, or green) go to the main disconnect. The neutral conductor (white or gray) goes directly to the neutral bus bar.

In addition to the three conductors, the drawing shows one other important wire associated with your service entrance panel — the grounding electrode conductor. This conductor, which connects the neutral bus bar to some permanently grounded object (usually a buried metal cold-water pipe), provides an electrical path to earth for your entire electrical system.

How to read your meter

Learning to read your electric meter can help you keep close track of your energy consumption, check your electricity bill, or simply satisfy your curiosity about that silent, sleepless meter.

Most electric meters have four or five dials with numbers and pointers. Most meters in older installations have four dials; meters installed in new service usually have five. A quick look at one of these dials shows that their numbering alternates between clockwise and counter-clockwise.

To take a reading of your meter, jot down the numbers indicated by the pointers, starting with the left dial. When the pointer is between two numbers, *always* read the smaller number. When a pointer appears to be directly on a number, check the next dial to the right. If the pointer of that dial is on zero or has passed zero, record the number indicated by the pointer of the first dial. If the pointer of the second dial has not yet reached zero, write down the next smaller number than the one indicated on the first dial.

In the illustration below, the pointer of the first (left-hand) dial is between 0 and 1, so the

KILOWATT HOURS

number we record is "0"; in the second dial, the pointer is between 4 and 5, so we write "4." The pointer of the third dial is almost directly at 2, so we look at the pointer of the fourth dial. It has passed 9 but not quite reached 0. For the third dial, then, "1" is the number to record; for the fourth dial the number is "9." And on the last dial, since the pointer is between 7 and 8, we write the number "7." The reading is therefore 04197 kilowatt-hours.

If you want to figure out the number of kilowatt-hours consumed during a certain period of time, subtract the meter reading at the start of the period from the reading at the end. You can use this to check your utility bills.

9-A: SERVICE ENTRANCE

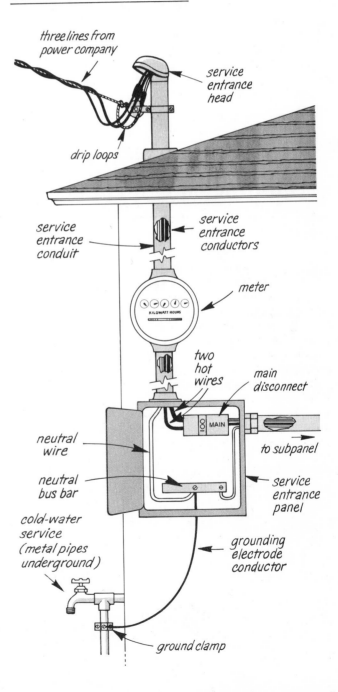

three lines from power company

service entrance head

drip loops

service entrance conduit

service entrance conductors

meter

two hot wires

main disconnect

to subpanel

neutral wire

neutral bus bar

service entrance panel

cold-water service (metal pipes underground)

grounding electrode conductor

ground clamp

LOCATION AND TYPE OF SERVICE EQUIPMENT

Though the ultimate capabilities are usually the same, the exact location and type of service equipment vary from home to home. As an example, consider the service entrance panel. It might be on the outside of your home, below the meter, or it might be on an inside wall, often directly behind the meter. It might have a single main disconnect, or it could have as many as six switches controlling disconnection (see "Shutting off main power supply," below). Also, the service entrance panel may or may not contain branch circuit overcurrent protection devices.

Variations also occur in the type of overcurrent protection devices. Some systems use circuit breakers; others use fuses.

Because of this variation, you shouldn't be concerned if our drawings don't look exactly like your particular service equipment. The principles of safety and protection should be the same regardless of the location and type of service equipment.

DISTRIBUTION CENTER

The next step in following the electrical path is to trace the power to the distribution center.

10-A: SERVICE ENTRANCE PANEL

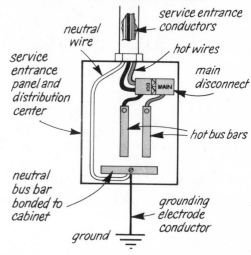

Shutting off main power supply

Most service entrance panels have a switch that enables you to disconnect your entire electrical supply instantly. This instant shut-off feature, also called the "main disconnect," is important whenever you work on existing wiring or make major repairs — or in case of an emergency, such as a fire.

Look at your main disconnect (usually identified as "Main") to see if it's one of the following types:

Lever disconnect. An external handle controls contact with two main fuses in the cabinet. When you pull the handle to the OFF position, you shut off the main power supply.

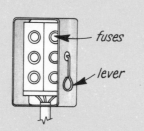

Single main circuit breaker. Switching the main breaker to the OFF position shuts off all power.

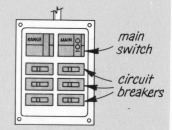

Pull-out block. The main cartridge fuses are mounted on one or two nonmetallic pull-out blocks. By pulling firmly on the hand-grips, you can remove the blocks from the cabinet and disconnect all power.

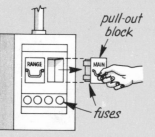

Multiple breaker main. Under the "rule of six" of the National Electrical Code, some homes are not required to have a single main disconnect. In such cases *all* of the breakers in the main section (not more than six) constitute the main, and *all* must be switched to OFF to disconnect all power.

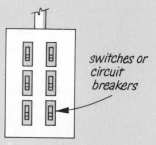

Drawings **10-A** and **11-A** show the two most common arrangements for the distribution center — note that it may be either part of the service entrance panel or a separate subpanel, located elsewhere in the home.

After passing through the main disconnect, each hot conductor connects to one of two hot bus bars in the distribution center. These bus bars accept the amount of current permitted by the main fuses or circuit breaker and allow you to divide that current into smaller units for the branch circuits.

Branch circuits. Each branch circuit attaches to one or both hot bus bars by means of an overcurrent protection device — a fuse or a circuit breaker. (For more information on the role of fuses and circuit breakers, see pages 20–21.)

Each 120-volt circuit consists of one hot conductor and one neutral conductor. The hot conductor originates at a branch circuit overcurrent protection device connected to one of the hot bus bars. Because a 240-volt circuit requires both hot conductors, it originates at a branch circuit overcurrent protection device connected to both hot bus bars.

All neutral conductors originate at a neutral bus bar in the distribution center. They are all in direct electrical contact with the earth through the grounding electrode conductor at the neutral bus bar of the service entrance panel. In order that ground potential (zero volts) be maintained at all

11-A: SUBPANEL DISTRIBUTION CENTER

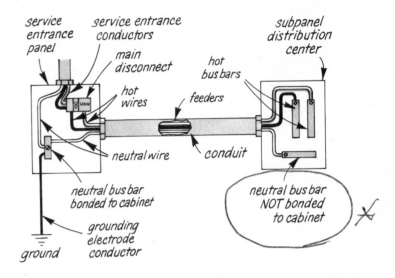

times, a neutral conductor must *never* be interrupted by an overcurrent protection device.

Drawing **11-B** shows distribution of electricity through a typical 100-amp service panel to the circuits of a kitchen. Note that each circuit takes off from an overcurrent protection device and returns to the neutral bus bar.

11-B: TYPICAL KITCHEN CIRCUITS

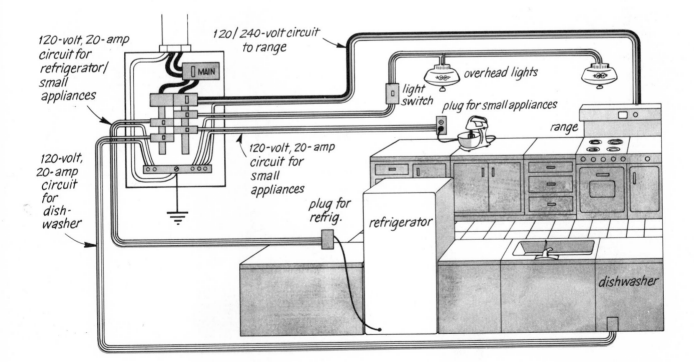

Simple circuitry

Now that we have a general understanding of the way a home electrical system works, let's take a closer look at the specific functions of the hot, neutral, and grounding wires. We will take a step-by-step approach, starting with a simple, partial circuit with just one light bulb and ending with drawings of typical arrangements for house circuits.

Simple light circuit

So far, we have learned that a circuit is a continuous closed path on which electricity flows from the source of voltage through a device using the electricity and back to the source. This principle is illustrated in drawing **12-A.**

12-A: SIMPLE CIRCUIT

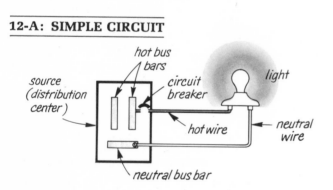

We call this a "partial circuit" because for the moment we are disregarding switches and equipment grounding wires.

Parallel wiring. In most homes several light fixtures operate on the same circuit by what we call "parallel wiring." With parallel wiring the hot and neutral wires run continuously from one fixture box to another. Wires to the individual lights branch off from these continuous hot and neutral wires.

Drawing **12-B** shows a partial circuit of lights wired in parallel.

Series wiring. Often contrasted to parallel wiring is series wiring. When lights are wired in series, the hot wire passes through all of the lamps before joining the neutral wire which returns to the source (see drawing **12-C).**

Series wiring is rarely used for home light

circuits because when one light bulb fails, all the lights go out. A string of old-style Christmas tree lights is an example of series wiring.

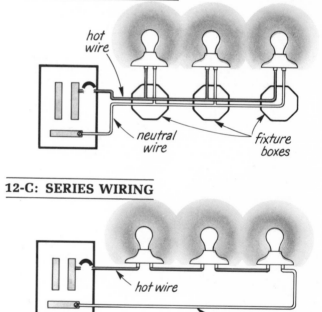

12-B: PARALLEL WIRING

12-C: SERIES WIRING

Light circuit with a switch

Our first step in building on the simple light circuit is to add a switch. Switches turn things on and off by controlling the flow of electric current. The simple knife-blade switch shown in drawings **13-A** and **13-B** illustrates how a switch closes (completes) and opens (breaks) a circuit, turning the light on and off.

13-A: CIRCUIT COMPLETE

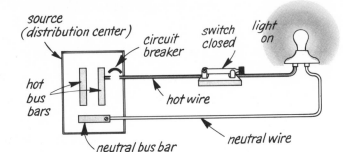

13-B: CIRCUIT BROKEN

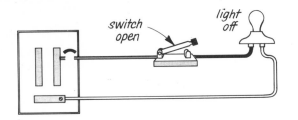

Though the switches in our homes don't look like the switch in the drawings, they work on the same principle. The important concept to remember is that *switches are installed only in hot wires.*

The switch in a hot wire disconnects the device from the hot bus bar as shown in drawing **13-B**. This leaves the device at ground potential (zero volts) and thus eliminates the possibility of a shock or short circuit at the device when the switch is open. A switch in the neutral wire would also interrupt the flow of current, but it would not disconnect the device from the hot bus bar. As a result, a shock or short circuit at the device would be possible.

13-C: LIGHT CONTROLLED BY SWITCH

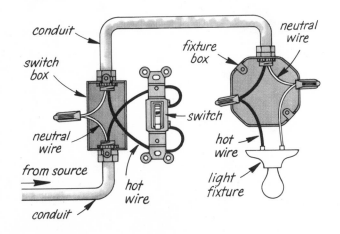

Drawing **13-C** is a more realistic view of part of the same circuit just discussed. It shows a light fixture controlled by a single-pole switch. Individual conductors were pulled through conduit for the wiring shown in this drawing. Once again, notice that the switch is in the hot wire; the neutral wire bypasses the switch and goes directly to the light fixture.

COLOR CODING—AND AN EXCEPTION

Up to this point we have assumed that a white wire is always a neutral wire. Wires that are black, red, or any color other than white, green, or gray are always hot. But one situation offers an exception to this color coding.

Drawing **13-D** shows the wiring of a switch loop with two-wire cable which is purchased with one black wire and one white wire. In this case, the white wire may substitute as the hot wire going from the source to the switch.

When using a white wire this special way, paint the insulation black (or tape it with black tape) at both ends where it joins a hot terminal or another hot wire. This identifies it as a hot wire. This exception to standard color coding is important because most home wiring today is done using cables.

13-D: SWITCH LOOP

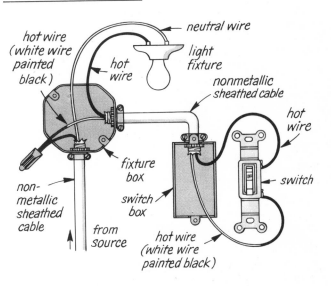

THREE-WAY SWITCHES

Drawings **13-C** and **13-D** show a single-pole switch that controls a fixture from only one location. But if you want to be able to turn a light on and off at two locations, such as at the top and the bottom of stairs, use a pair of three-way switches.

Though single-pole and three-way switches look somewhat alike, two features distinguish them. A single-pole switch has two terminals for wire connections and the words "ON" and "OFF" embossed on the toggle. A three-way switch, on the other hand, has three terminals; "ON" and "OFF" are not indicated on the toggle, since the on and off positions may change, depending on the position of the other switch.

In order to clarify the action of three-way switches, let's put a pair of them in a simple light circuit. For illustration, we show each switch as three terminals and a movable blade.

14-A: THREE-WAY SWITCHES

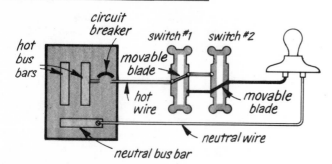

Light is off when circuit is open, one switch up and one down (**14-A**). Flip either switch so both are up or down (**14-B**), circuit is completed; light is on.

14-B

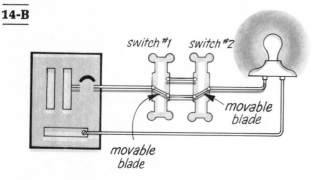

Grounding

Up to this point our discussion of circuits has disregarded grounding. Instead we have worked with just two wires — a hot and a neutral. Electrical codes now require that every 120-volt circuit have a system of grounding. This is a preventive measure, much like oxygen masks on a commercial airliner. During normal operation the grounding system does nothing; in the event of a malfunction, however, the grounding is there for your protection.

Grounding assures that all metal parts of a circuit that you might come in contact with are maintained at zero voltage because they are connected directly to the earth.

To see why grounding is necessary, let's start by looking again at the simple circuit in drawing **12-A**, which shows a circuit during normal conditions. Now let's take that same circuit and add a metal pull-chain switch. If the hot wire accidentally became dislodged from the fixture terminal and came into contact with the metal canopy of the light fixture, the fixture and pull chain would become electrically charged, or "hot." If you were to touch the chain or fixture under these conditions, a short circuit could occur in which you would provide the path to ground for the electric current. In other words, you would get a shock (see drawing **14-C**).

This same situation could occur in any number of places where electricity and conductive materials are together — in power tools and appliances with metal housings; in metal switch, junction, and outlet boxes; and in metal faceplates.

The shock in our example could have been prevented if the circuit had a grounding system. A grounding wire connecting the neutral bus bar to the metal housing of the light fixture would provide an auxiliary electrical path to ground in the event of a short circuit. This grounding wire would carry the fault current back to the distribution center and assure that the fuse or circuit breaker protecting the circuit would open, shutting off all current flow.

14-C　　　　　　　　**14-D: GROUNDING PREVENTS SHOCK**

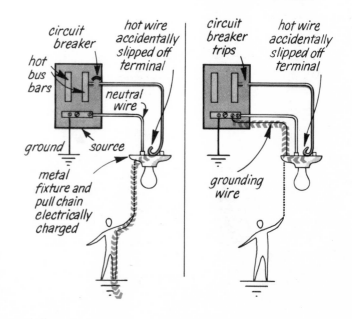

GROUNDING YOUR ELECTRICAL SYSTEM

Shifting our attention now to a typical house circuit, we will see that the wiring method dictates how grounding is done.

When a home is wired with armored cable, metal conduit, or flexible metal conduit (in which both conduit and fittings are approved for the purpose by your local electrical code), the conductive metal enclosures can themselves form the grounding system as shown in drawing **15-A**. When metal enclosures are not used, a separate grounding wire must be run with the circuit wires. Running a separate grounding wire isn't as complicated as it may sound because nonmetallic sheathed cable is available with a bare grounding wire already in the assembly (see page 37).

In both systems the end result is the same: an auxiliary path for fault current is provided to the neutral bus bar in the service entrance panel, which is tied to ground.

In drawing **15-A**, grounding continuity is maintained through metal enclosures that are bonded together. The final grounding connection to the receptacle is made through a short piece of wire (called a "jumper") that is bonded to the box with either a grounding screw or a grounding clamp.

15-A: GROUNDING WITH CONDUIT

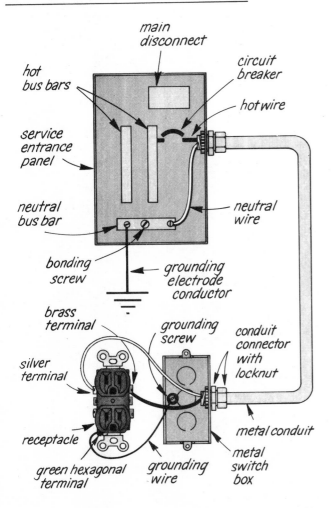

15-B: GROUNDING WITH NM CABLE

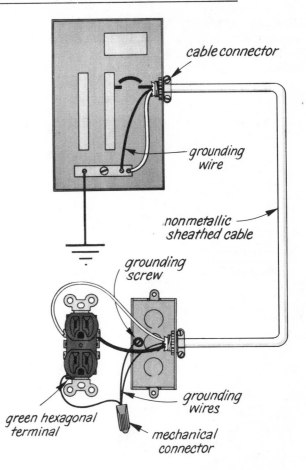

In drawing **15-B**, the bare grounding wire of the nonmetallic sheathed cable provides the grounding connection. The receptacle and metal box are grounded with jumpers that are held together with the grounding wire of the cable by a mechanical connector. If a nonmetallic box were used instead, the grounding wire would be connected directly to the receptacle because the box needs no grounding.

Circuit diagrams

Drawings **16-A** to **19-C** show some typical arrangements for switches, fixtures, and receptacles using nonmetallic sheathed cable with ground, metal boxes, and twist-on connectors for wire splices. All of the light fixtures are grounded to the boxes through the mounting screws. Many other combinations are possible of course, but these drawings should guide you in planning your circuit arrangements.

Before doing any wiring, check your local code to find out if there are any specific equipment requirements.

16-A: LIGHT AT END OF CIRCUIT

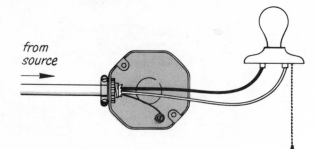

Circuit ends at light fixture with pull chain switch.

16-B: LIGHT AT END OF CIRCUIT

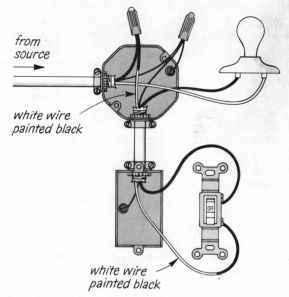

from source

white wire painted black

white wire painted black

Single-pole switch controls light at end of circuit.

16-C: LIGHT AT END OF CIRCUIT

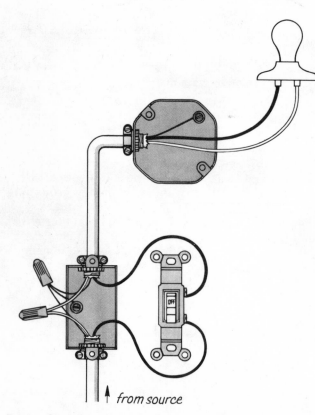

from source

Power from source goes through single-pole switch that controls light at end of circuit.

16-D: LIGHT IN MIDDLE OF CIRCUIT

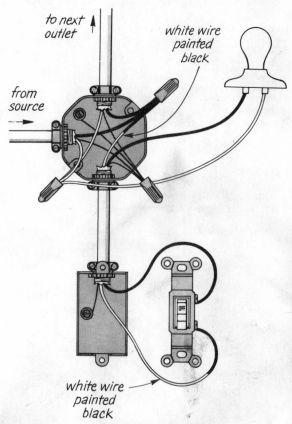

to next outlet

white wire painted black

from source

white wire painted black

Switch wired in switch loop controls light in middle of circuit run.

17-A: RECEPTACLE AT END OF CIRCUIT

17-B: THREE-WAY SWITCHES

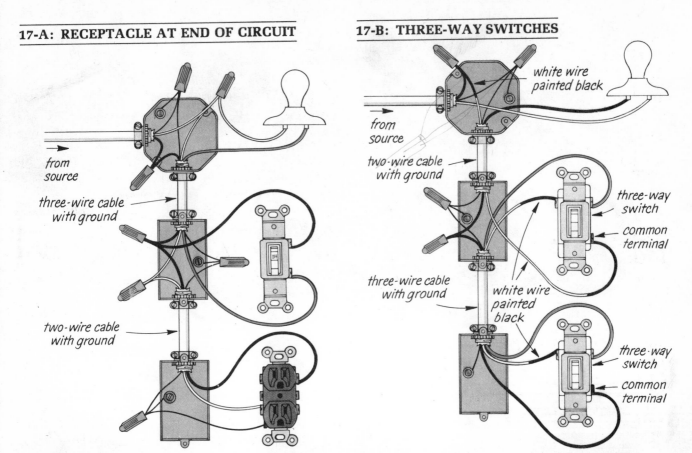

Switch wired ahead of receptacle in circuit. Switch controls light; receptacle is always hot.

Light at end of circuit is controlled by a pair of three-way switches.

17-C: THREE-WAY SWITCHES

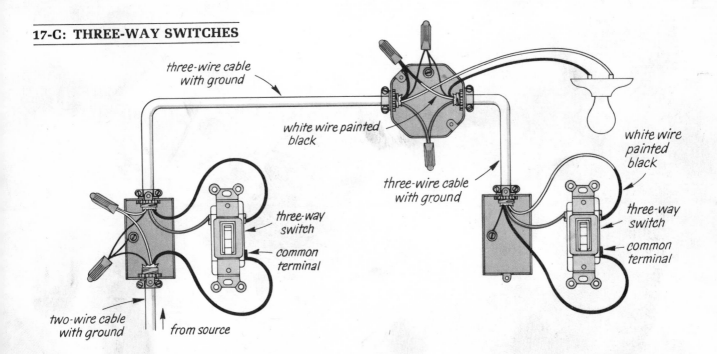

Light wired between a pair of three-way switches at end of circuit.

18-A: THREE-WAY SWITCHES

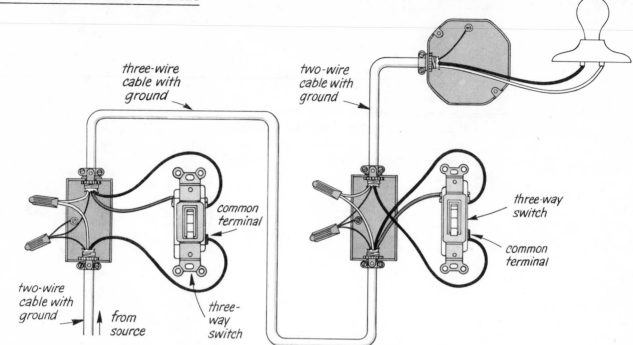

three-wire
cable with
ground

two-wire
cable with
ground

common
terminal

three-way
switch

common
terminal

two-wire
cable with
ground

from
source

three-
way
switch

Power goes through *pair of three-way switches to light at end of circuit.*

18-B: THREE-WAY SWITCHES AND HOT RECEPTACLE

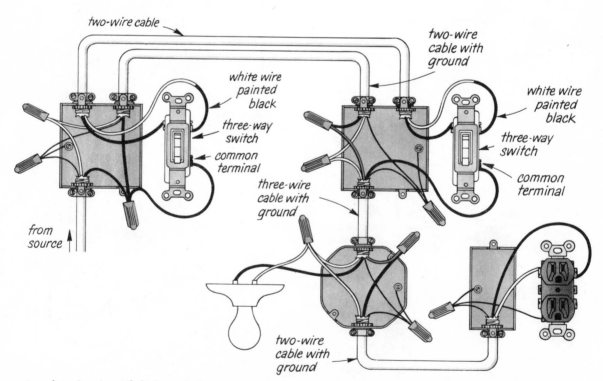

two-wire cable

two-wire
cable with
ground

white wire
painted
black

three-way
switch

common
terminal

white wire
painted
black

three-way
switch

common
terminal

three-wire
cable with
ground

from
source

two-wire
cable with
ground

One way to wire *circuit with light and three-way switches so receptacle at end is always hot.*

18 Simple circuitry

19-A: RECEPTACLES AT END OF CIRCUIT

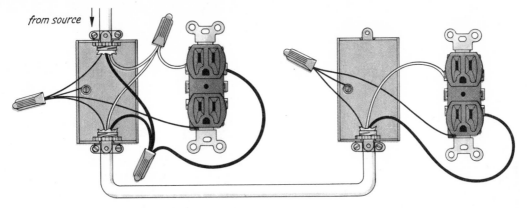

from source

Wired in parallel, *two receptacles at end of circuit are always hot.*

19-B: SWITCH-CONTROLLED RECEPTACLE

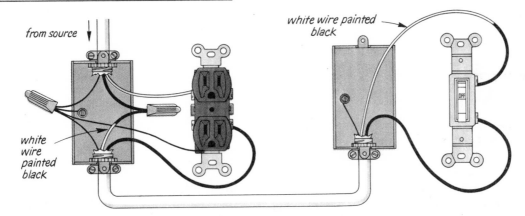

from source

white wire painted black

white wire painted black

Receptacle at end *of circuit is controlled by single-pole switch.*

19-C: SPLIT-CIRCUIT RECEPTACLE

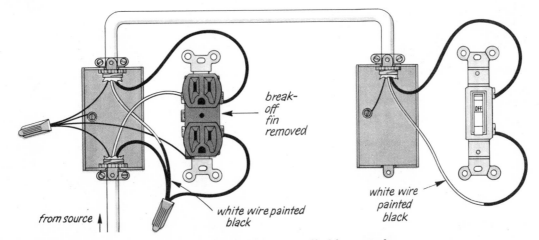

break-off fin removed

from source

white wire painted black

white wire painted black

Half of duplex receptacle *is always hot, and other half is controlled by switch.*

Simple circuitry **19**

Do-it-yourself maintenance & repair

You could expect quite a charge—financially, that is—if you had to call in an electrician every time the lights went out. This chapter will help you toward electrical independence by explaining what a blown fuse and a tripped circuit breaker are and how you can deal with them. Instructions for repairing lamps and doorbells will also help you to be self-sufficient and thrifty.

Safeguards in your electrical system

Fuses and circuit breakers, collectively referred to as "overcurrent protection devices," guard electrical systems from damage by too much current.

Whenever wiring is forced to carry more current than it can safely handle — whether it's because of a sudden surge from the power company, use of too many appliances on one circuit, or a problem within your system — fuses will blow or circuit breakers will trip. These actions open the circuits, disconnecting the supply of electricity.

A circuit breaker or fuse is inserted into each circuit at the service entrance panel (or in some cases at a subpanel). For adequate protection, the amperage rating of a breaker or fuse must be the same as that of the circuit conductor it protects. For example, a circuit using #12 copper conductor has an ampacity of 20 amps (see **Table IV**, page 42); the fuse or circuit breaker, therefore, must also be rated for 20 amps. *Never replace any fuse or circuit breaker with one of higher amperage.*

FUSES

A fuse contains a short strip of an alloy with a low melting point. When installed in a socket or fuseholder, the metal strip becomes a link in the circuit. If the amperage flowing in the circuit becomes greater than the rating of the fuse, the metal strip will melt, opening the circuit.

top view of fuses

good fuse   blown fuse

Edison base fuses. Equipped with screw-in bases like those of ordinary light bulbs, Edison base fuses come in ratings up to 30 amps. According to the National Electrical Code, Edison base fuses are now permitted only as replacements in 120-volt circuits. (This Code restriction is meant to encourage everyone to use the safer, nontamperable Type "S" fuse, described below.)

Edison base fuse (side view)

Type "S" fuses. You must install the adapter with the correct rating in the fuse socket before using a Type "S" fuse. Each adapter is constructed so that it is impossible to install a fuse with a higher rating. Fuses can be replaced as needed, but once an adapter is installed, it can't be removed. Type "S" fuses are required in all new installations that use fuses to protect 120-volt circuits.

Type "S" fuse

Cartridge fuses. There are two basic styles of cartridge fuses: ferrule and knife-blade.

Ferrule type fuses, which come in ratings of 10 to 60 amps, are usually used to protect the circuit of an individual 120/240-volt appliance, such as a range.

Available in ratings of 70 amps or more and suitable for 240 volts, knife-blade fuses are generally used as the main disconnect in fused service entrance panels.

(A fuse pulling tool is best for removing cartridge fuses from their fuseholders.)

cartridge fuses

ferrule type

knife-blade type

CIRCUIT BREAKERS

Resembling a light switch, a circuit breaker serves both as a switch and as a fuse. As a switch, a circuit breaker lets you open a circuit (turn switch to OFF) whenever you want to work on the wiring. As a fuse, it provides automatic overcurrent protection.

When a breaker is installed in a circuit breaker panel, a bimetallic strip becomes a link in the circuit. Heat from excessive current will bend the metal strip, causing a release to trip and break the circuit. (The toggle goes to OFF or to an intermediate position when this happens.)

Unlike fuses, which work on the self-destruct principle, circuit breakers can be reset (turned back on) once they've tripped.

All circuit breakers are rated for a specific amperage. As with fuses, the amp rating of a breaker must match the ampacity of the circuit it protects.

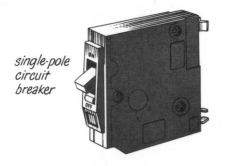

single-pole circuit breaker

What if the lights go out?

What do you do if a light doesn't go on when you flip the switch? Most likely the cause is a burned out light bulb, but first make a quick check to see if another light on the same circuit works. If it does, replace the bulb with a new one.

If the second lamp doesn't light, suspect a dead circuit. Check your service panel for a blown fuse or tripped circuit breaker. A blown fuse or tripped breaker tells you one of two things. You have either a short circuit or an overloaded circuit.

If your entire home is without electricity and there are no blown fuses or tripped circuit breakers, check neighboring homes. If they seem to be in the dark too, your area probably has a local power failure. A phone call to your utility company will usually bring prompt attention to the problem.

If your home is the only one without electricity, recheck your service entrance panel to see if the main fuse has blown or the main circuit breaker has tripped. If the main is intact, call your utility company. You may have a downed line.

⚠ Stay away from power lines—up or downed.

Overloaded circuit. If you're in partial darkness, the first possibility to consider is an overloaded circuit. Without realizing it, you may have put too many lights and appliances on a circuit.

As you plug in appliances, the current increases to meet the demands. When the current exceeds the safe amperage limits of the circuit, the fuse will blow or the circuit breaker will trip.

If a circuit goes dead as soon as you turn on an electric iron while dinner is cooking in the microwave oven, your loss of electricity probably results from an overloaded circuit. Unplug the iron and then replace the fuse or reset the circuit breaker. This should restore your electricity — but it won't get the ironing done. To take care of this, plug some appliances from the overloaded circuit into other circuits. This redistribution should allow you to finish your ironing with no further problems.

Short circuit. When an accidental path is created between a hot wire and any ground, lots of current will flow. This situation is called a "short circuit." The name comes from the fact that it provides a shorter path to ground than the intended circuit.

An exposed wire spells trouble, whether it's exposed because of worn insulation or a faulty connection. All that has to happen is for the exposed area to touch any grounded object (such as the neutral wire, a grounding wire, a grounded metal box, or grounded metal conduit) and you'll have a short circuit. This is a frequent problem with two-wire cord such as zip cord. A short circuit will occur whenever the two wires touch each other, since one wire is hot and the other is grounded (neutral).

How to replace a screw-in fuse

Keep spare fuses handy — you might, for instance, tape them to the door of your fuse cabinet. Make sure that your extra fuses have the appropriate amperages for your circuitry.

When replacing a screw-in fuse, keep these shock prevention measures in mind:
• Grasp the fuse only by its outer glass rim.
• Never stand on a damp spot. Put down a dry board.
• So you'll be less likely to form a complete circuit, work with only one hand at a time when possible. (It's a good idea to keep your free hand behind your back or in a pocket to resist the temptation to assist the other hand.)

In addition, give Type "S" fuses an extra turn or two to assure good contact with the fuseholder.

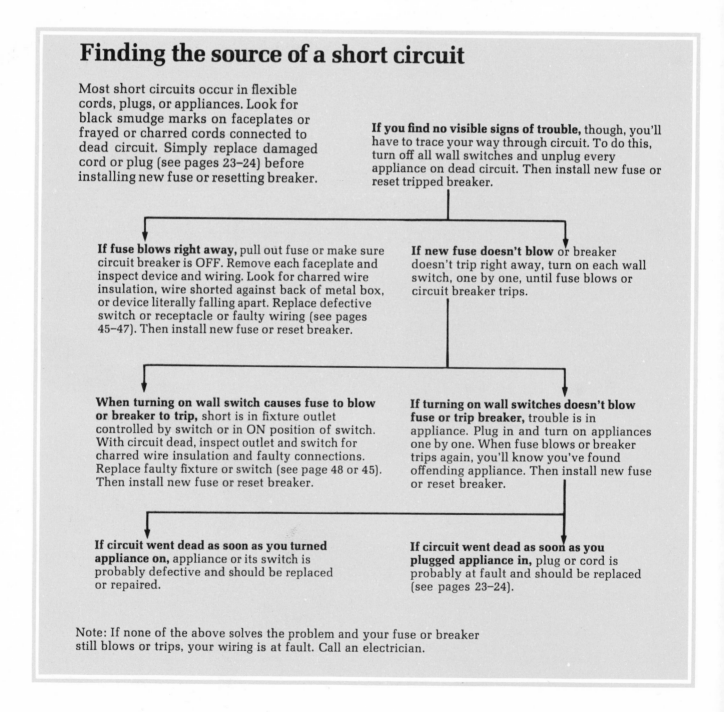

Finding the source of a short circuit

Most short circuits occur in flexible cords, plugs, or appliances. Look for black smudge marks on faceplates or frayed or charred cords connected to dead circuit. Simply replace damaged cord or plug (see pages 23–24) before installing new fuse or resetting breaker.

If you find no visible signs of trouble, though, you'll have to trace your way through circuit. To do this, turn off all wall switches and unplug every appliance on dead circuit. Then install new fuse or reset tripped breaker.

If fuse blows right away, pull out fuse or make sure circuit breaker is OFF. Remove each faceplate and inspect device and wiring. Look for charred wire insulation, wire shorted against back of metal box, or device literally falling apart. Replace defective switch or receptacle or faulty wiring (see pages 45–47). Then install new fuse or reset breaker.

If new fuse doesn't blow or breaker doesn't trip right away, turn on each wall switch, one by one, until fuse blows or circuit breaker trips.

When turning on wall switch causes fuse to blow or breaker to trip, short is in fixture outlet controlled by switch or in ON position of switch. With circuit dead, inspect outlet and switch for charred wire insulation and faulty connections. Replace faulty fixture or switch (see page 48 or 45). Then install new fuse or reset breaker.

If turning on wall switches doesn't blow fuse or trip breaker, trouble is in appliance. Plug in and turn on appliances one by one. When fuse blows or breaker trips again, you'll know you've found offending appliance. Then install new fuse or reset breaker.

If circuit went dead as soon as you turned appliance on, appliance or its switch is probably defective and should be replaced or repaired.

If circuit went dead as soon as you plugged appliance in, plug or cord is probably at fault and should be replaced (see pages 23–24).

Note: If none of the above solves the problem and your fuse or breaker still blows or trips, your wiring is at fault. Call an electrician.

⚠ Always match fuse size to circuit rating. For instance, if the circuit is rated for 15 amps, use a 15 amp fuse, *not* a 20 amp one.

How to reset a circuit breaker

The procedure for resetting a circuit breaker varies from brand to brand. Directions are often embossed on the breaker.

Many modern circuit breakers go to an intermediate position between ON and OFF when they trip. To reset the trip mechanism, push the toggle firmly to OFF before returning to ON. (Note: switching a circuit breaker may require more force than an ordinary household switch.)

Simple repairs

You don't need the know-how of an expert to keep your home in good electrical repair. You can replace plugs and light fixtures and repair lamps, fluorescent fixtures, and doorbells. Many repairs and improvements should be done as preventive maintenance. Don't wait until a defect causes a fire or short circuit.

Not everything can — or should — be repaired, however. For instance, a cord with badly frayed or brittle insulation is a hazard. It should be thrown away and *replaced*, not repaired.

See pages 45–48 if your repairs involve changing a switch, receptacle, or wall or ceiling-mounted light fixture.

REPLACING CORDS AND PLUGS

Danger signs on cords and plugs include arcing electricity, irregularly transmitted electricity, physical damage, and excessive heat (a cord or plug that's too warm to touch). Treat these symptoms promptly by replacing the defective plug.

Plugs are of two kinds: those that are self-connecting and those that have screw terminals. Drawing **23-A** shows how to attach a typical self-connecting plug.

23-A: SELF-CONNECTING PLUG

Step 1. *Use only zip cord with self-connecting plugs; cut end square.*

Step 2. *Pull out prongs from plug and push cord in as far as it will go.*

Step 3. *Squeeze prongs together and push them back into cover.*

Drawing **23-B** shows how to wire a two-prong plug with terminal screws. The same procedure is used for three-prong plugs (drawing **24-A**) and appliance plugs (drawing **24-B**).

23-B: SCREW TERMINAL PLUG

zip cord
inner insulation
insulator
outer insulation
plug
cord
Underwriters' knot
½ to ¾ inch

Step 1. Part zip cord 2 inches

... or remove 2 inches of outer insulation.

Step 2. Remove insulator and slip cord through plug.

Step 3. Tie Underwriters' knot in cord.

Step 4. *Strip off ½ to ¾ inch insulation from end of each wire and loosen terminal screws (don't force screws all the way out).*

Step 5. *Twist each stranded wire clockwise, then wrap one wire clockwise around each terminal screw. Tighten terminal screws and replace insulator.*

24-A: THREE-PRONG PLUG

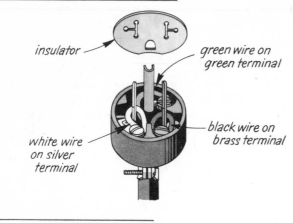

insulator

green wire on green terminal

white wire on silver terminal

black wire on brass terminal

24-B: APPLIANCE PLUG

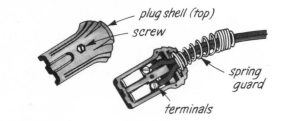

plug shell (top)

screw

spring guard

terminals

24-C: ANATOMY OF A LAMP

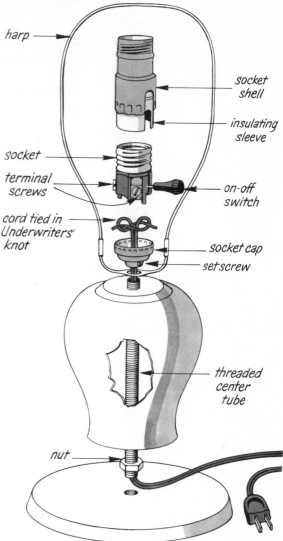

harp

socket shell

insulating sleeve

socket

terminal screws

on-off switch

cord tied in Underwriters' knot

socket cap

set screw

threaded center tube

nut

Knowing where and how *the parts of a lamp fit together is useful when you wish to replace a socket, rewire a lamp, or build a lamp from scratch.*

REPAIRING LAMPS

Most plug-in incandescent lamps are electrically alike. Table or floor model, large or small, straight or curved, wood or brass, they all have a socket, switch, cord, and plug. And these are the four elements that may wear out and cause your lamp not to work. Replacing the defective part is usually all that is necessary to restore the lamp.

The assembly of some lamps, though, prevents their repair. If the faulty lamp was assembled with rivets instead of nuts and bolts, the lamp is a better candidate for the junkpile than the repair bench.

Troubleshooting a lamp

If your lamp doesn't work, diagnose the trouble by taking the following three steps:
1) Check the light bulb to make sure it is a good one and check that it is screwed in as far as it will go. Replace the bulb if necessary.
2) Check the cord and plug for breaks and frayed areas. Replace if necessary (see above and page 23).
3) If the bulb, cord, and plug are good, replace the socket.

How to replace a lamp socket

Be sure to unplug your lamp before doing any work on it. Remove the old socket, leaving the Underwriters' knot and socket cap in place. To do this,

first remove the socket shell by squeezing and lifting near the switch, where the word *press* is embossed. Remove the insulating sleeve, then loosen terminal screws and disconnect the cord. It's a good idea to take the old socket along when you buy a new one to assure that you get a proper replacement.

The new socket you buy will include a socket cap, but there is often no need to replace the existing one. Loop the wires clockwise around the terminal screws of the new socket and tighten the screws. Place the insulating sleeve over the socket and then put on the socket shell. Make sure that the corrugated edges of the shell fit inside the rim of the cap, and then push them together until you hear the shell click into place.

How to rewire a lamp

Take a close look at your lamp before you remove the old cord completely. If the cord twists or curves inside the lamp, you might want to use the following method of feeding the new cord through:

1) Remove the socket shell, insulating sleeve, and wires from the terminal screws, then untie the Underwriters' knot.

2) Attach the new cord to the old cord by stripping off about ¾ inch of insulation from the new cord and twisting the wires of the two cords together. A few turns of tape over the twisted wires will strengthen the joint and make it less likely to catch on its way out.

3) Pulling carefully on the old cord, thread the new cord through the lamp. When the new cord is in place, detach the old cord.

Connecting the new cord to the socket is a process very similar to replacing a plug (see drawing **25-A**).

Reassemble the socket as described previously under "How to replace a lamp socket." To attach a

25-A: LAMP SOCKET

Step 1. *Thread cord through socket cap and part zip cord 2½ inches or remove 2½ inches of outer insulation.*

Step 2. *Tie Underwriters' knot in cord.*

Step 3. *Strip off ½ to ¾ inch of insulation from the end of each wire; loosen terminal screws, but don't force screws all the way out.*

Step 4. *Tightly twist each stranded wire clockwise. Wrap one wire clockwise around each terminal screw and tighten the screws.*

plug to the other end of the new cord, see directions on page 23.

FIXING FLUORESCENT LAMPS

Bulb for bulb and watt for watt, a fluorescent light provides more light for your money than an incandescent light does. For example, a 40-watt fluorescent tube produces almost six times as much light as a 40-watt incandescent bulb. And the fluorescent tube will last about five times as long as the incandescent bulb.

Unlike the simple principle of an incandescent bulb, which glows when current flows through the filament, an intricate electrical process takes place before a fluorescent tube gives light.

Because a fluorescent tube doesn't have a filament, a ballast (transformer) is necessary to set up voltage within the tube. In addition to a ballast, older style fluorescent fixtures also have a starter that assists the ballast in the initial starting process.

The two most common types of fluorescent light fixtures for homes are rapid-start and preheat. It is easy to distinguish between them because the starter mechanism of the rapid-start type is built right into the ballast; on the preheat type, each tube has a visible starter unit (see drawing **27-A**). The starters, which look like small aluminum cylinders, tend to burn out as often as the bulbs do.

A third type, less commonly used in the home, is the instant-start. This type has no starter and is distinguished by a tube with a single pin on each end.

Tubes, starters, ballasts, and tubeholders are the components usually involved in fluorescent lamp repair. All components are easy to replace, making most repairs a matter of substitution.

⚠ Make sure current is off before working on wiring.

How to replace the parts

Replacement parts for a fluorescent lamp must be carefully matched to the fixture. Tubes and ballasts can't be interchanged between the types of fixtures, and the starter (if there is one) and ballast must match the wattage of the tube. You'll find most of the information you need printed on the parts themselves.

Before working on any electrical appliance, you should disconnect it from the voltage source. With a table-model fluorescent lamp, this means unplugging the lamp. If your fixture is wall or ceiling mounted, you must de-energize the circuit by removing the fuse or tripping the circuit breaker.

Fluorescent tubes. To remove a double pin type fluorescent tube, twist it a quarter turn in either direction and gently pull out. Install a new tube by pushing it into the tubeholders and then giving it a quarter turn to lock it in place.

A single pin type tube is removed by pushing it against the spring-loaded tubeholder until the other end can be removed. To replace a tube, put the tube pin in the spring-loaded tubeholder and push until the other end can be inserted.

Starters. To get to the starter, remove the fluorescent tube. Remove the starter by twisting it a quarter turn counterclockwise and then pulling it out of its socket. Place a new starter in the socket and twist it clockwise for a quarter turn.

Tubeholders. Because there is considerable variety in tubeholders (sometimes called "sockets"), it's wise to take the one you're replacing with you when buying a new one.

To remove a damaged tubeholder, disconnect the wires first. If the wires are connected by terminal screws, loosen the screws to free the wires. If the tubeholder has push-in wire connectors, release each wire by inserting a small screwdriver or nail

Troubleshooting a fluorescent tube

SYMPTOM	CAUSE	CURE
Lamp won't light	Tube burned out (blackened ends)	Replace tube
	Improper installation	Take out and install again
	Fuse blown or circuit breaker tripped	Replace or reset
	Starter burned out	Replace starter
	Dirty tube (rapid-start only)	Remove tube, wash, rinse, dry, and replace
	Tubeholder broken	Replace tubeholder
	Fixture too cold	Raise temperature to at least 50°F
	Oxide film buildup on tube pins	Rotate tube in tubeholders once or twice
Lamp flickers (Note: New tubes may flicker a short time after installation.)	Poor contact with tubeholders	Realign tubeholders; straighten and sand tubeholders if necessary
	Improper installation	Take out and install again
	Tube nearly worn out (blackened ends)	Replace tube
	Oxide buildup on tube pins	Rotate tube in tubeholders once or twice
	Fixture too cold	Raise temperature to at least 50°F
Ends of tube are discolored (Note: Darkened bands about 2 inches from ends are normal.)	Tube almost worn out	Replace tube
If preheat type with new tubes	Defective starter	Replace starter
Discolored on one end only	Temperamental tube	Remove tube; turn end for end
Ends of tube glow, but center doesn't	Defective starter	Replace starter
	Defective ballast	Replace ballast
Lamp fixture hums	Ballast incorrectly installed	Check wiring on ballast diagram and correct
	Wrong type of ballast	Check wattage and type; replace ballast
	Defective ballast	Replace ballast

27-A: PREHEAT FIXTURE

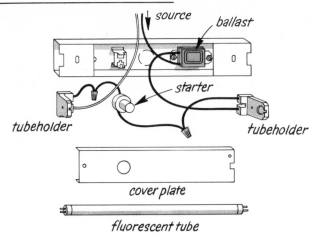

into the slot next to the connection. Then remove the tubeholder by taking out the mounting screw. (On some models you may have to take the end bracket off the fixture so you can slide off the tubeholder.)

Install the new tubeholder by reversing the preceding procedure — mount the holder, then connect the wires.

Ballasts. Shut off all current in the lamp. Disassemble the fixture until you get to the ballast. In table models, it is located in the base of the lamp; in ceiling or wall-mounted models, the ballast is in the metal enclosure that's attached to the ceiling or wall.

The replacement ballast you buy will already be wired, and about 5 inches of each wire will be sticking out of it. Remove the old ballast by cutting all the wires coming out from it about 5 inches from the ballast and undoing the mounting screws. Mount the new ballast before you connect its wires to the wiring in the fixture.

To connect the wires, strip about ½ inch insulation from the end of each wire and then use a medium-size wirenut (see page 44) for every two wires. When working with a ballast, match up the color-coded wires correctly: connect red to red, blue to blue, etc. Check your work against the wiring diagram printed on the ballast.

CURING SILENT DOORBELLS

Whether they ring, buzz, or chime, all doorbell systems operate with low voltage — that is, voltage significantly lower than the 120 volts of normal household current. Drawing **27-B** shows how a transformer is wired into a doorbell circuit to step down the 120-volt current to anywhere from 6 to 24 volts, depending on the capacity of the transformer and bell.

For most diagnostic work on your doorbell, you must have the power source connected. However, always de-energize the circuit by pulling the fuse or tripping the breaker if you are going to work on the transformer. (Remember, the input side of the transformer is high voltage — 120 volts.)

The first thing to examine is the source of power. Make sure that the fuse or circuit breaker protecting the circuit hasn't blown or tripped. Once you are assured that the 120-volt side of the transformer is getting power, go on to check the low-voltage side.

While someone else goes around to the door and pushes the button, listen to the transformer. If you hear a humming sound, indicating that the transformer is working, the problem is elsewhere — possibly in the bell mechanism. Follow the advice in Part A below. If you don't hear a hum, your transformer could be defective or there could be a break somewhere in the circuit. In that case go to Part B.

27-B: DOORBELL CIRCUIT

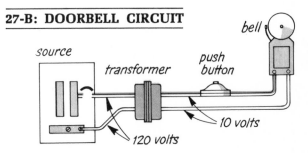

Part A. Occasionally a doorbell won't ring because it is gummed up with dirt and grime. For example, the striker shaft on a chime mechanism can become immobilized by corrosion or grease. Check the mechanism and clean as necessary.

If the bell mechanism still does not work, proceed with Part B to check for a poor or broken connection.

Part B. The first step is to inspect the wiring. Because low-voltage wiring uses such small wire (usually #18 for doorbells), the wires sometimes break, fray, or lose their insulation. Look for signs of wear; tape the wires as necessary.

Next, check wire connections carefully at the transformer, bell mechanism, and push button. Using fine sandpaper and a screwdriver, remove corrosion from contacts; clean and tighten terminals as necessary. This may require that you remove the button from the wall.

To check that the button is working, disconnect the two wires and touch the bare ends together. If this makes the bell ring, the push button is the culprit and should be replaced.

Taping wires, cleaning contacts, and tightening connectors are often all that's necessary to give a voice to a silent doorbell. Try ringing yours; chances are it will work now.

Mapping out your wiring plan

"What do I have to work with?" is the first question you should ask yourself when you consider any repairs, alterations, or additions to your present electrical system. It's a lot like asking yourself how much money you have in the bank before going out to buy something expensive. In this chapter we'll help you evaluate your electrical assets and plan for future modifications.

Evaluating your electrical assets

The first step in evaluating your electrical system is to determine what type of electrical service you have. Looking through the glass window of your electricity meter, you'll see several numbers on the faceplate. "120V" indicates two-wire service; "240V" indicates three-wire service with both 120-volt and 240-volt capabilities.

SERVICE RATINGS

Any electrical system is rated for a maximum amount of current (measured in amperes) it can carry. This rating, determined by the size of the service entrance equipment, is called the "service rating." Keeping you within the bounds of your service rating is the job of the main fuses or circuit breaker.

Service ratings have increased through the years to accommodate greater electrical demands and higher safety standards. Today, the minimum service rating of most new homes is 100 amps. Depending to a large extent on the age of your home, your service rating could be as low as 30 amps or as high as 400 amps. In between the two extremes are the following common service ratings: 60, 100, 125, 150, and 200 amps.

The best way to find out your service rating is to look at the main disconnect, if you have one. Whether it is a breaker or fuses (see page 10), the service rating will usually be stamped on the main fuses or circuit breaker.

If your system doesn't have a main disconnect, call the utility company or your local building inspection department rather than trying to figure out the service rating. Someone from either of those two offices will be able to advise you about the rating of your service.

KNOW YOUR CIRCUIT STRUCTURE

Before you let yourself daydream about electrical modifications to your home, you should know which fuse or circuit breaker protects what receptacles, light fixtures, and switches. A basic wiring diagram of your entire house is one of the most useful aids you can have.

To decode your wiring, start by giving a number to each fuse or breaker in the distribution center (if you have more than one subpanel, be sure to number all branch circuits). Next, draw a map showing each room, including the basement and garage. Using the symbols shown in "Circuit mapping" page 29, indicate on the map approximate locations of each receptacle, fixture, and switch.

Chart the 120-volt circuits first. To do this, you will need a small table lamp or night light that you can easily carry around with you to test all of the receptacles.

Turning one breaker to the OFF position (or removing one fuse), systematically go through the house and check all switches and receptacles. On your map write the circuit number (the number assigned to the overcurrent protection device) next

(Continued on page 30)

Circuit mapping

Using numbers and electrical symbols, you can make up a good working drawing of your electrical system. Such a drawing or map can save you much time, whether you plan to wire a new home, alter existing wiring, or trouble-shoot a problem.

The following is a circuit map of a typical two-bedroom house. Note that the dashed lines indicate what switch controls what fixture; they do not show wire routes.

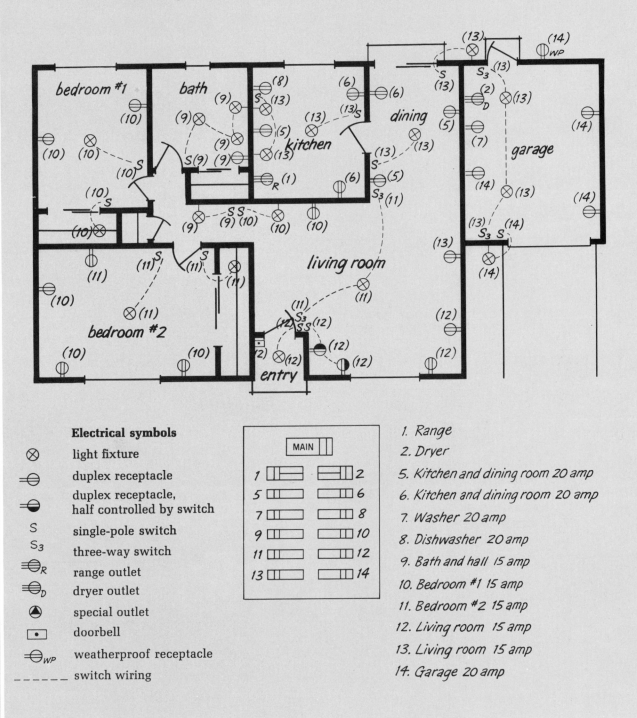

Electrical symbols

⊗ light fixture

⊖ duplex receptacle

⊖ duplex receptacle, half controlled by switch

S single-pole switch

S₃ three-way switch

⊖R range outlet

⊖D dryer outlet

⬤ special outlet

▭ doorbell

⊖wp weatherproof receptacle

------ switch wiring

1. Range
2. Dryer
5. Kitchen and dining room 20 amp
6. Kitchen and dining room 20 amp
7. Washer 20 amp
8. Dishwasher 20 amp
9. Bath and hall 15 amp
10. Bedroom #1 15 amp
11. Bedroom #2 15 amp
12. Living room 15 amp
13. Living room 15 amp
14. Garage 20 amp

. . . Continued from page 28

to each switch, fixture, and receptacle that is now dead. After you've made a complete check of the house, turn the breaker back on or replace the fuse, and start the process over with another circuit.

Here are several points to remember when testing 120-volt circuits:
• Make sure the lamp you use for testing works and is turned on.
• Make sure to test both outlets of a duplex receptacle.
• Don't forget to test the garbage disposal switch and dishwasher.

Once the 120-volt circuits are charted, go on to the 240-volt circuits. These circuits — identified in your distribution center by a double circuit breaker or a pull-out block with cartridge fuses — go to individual high-wattage appliances such as an electric range, clothes dryer, water heater, heating system, or central air conditioner. Trace the 240-volt circuits by disengaging one overcurrent protection device at a time and finding which appliance doesn't work.

CALCULATE YOUR ELECTRICAL USAGE

Once you know what your electrical assets are, the next step is to determine your present usage, or electrical load. This would be a time-consuming task if you had to go around the house and add up all the wattages of the lights and appliances. However, after considerable research the National Electrical Code has established certain values that represent typical electrical usage.

Three watts per square foot (using outside dimensions) of living space and space for future use is the figure used for general purpose circuits (general lighting and receptacles). A nominal value of 1,500 watts is used for each 20-amp small appliance circuit (circuits that power receptacles in the kitchen, dining room, family room, breakfast room, and pantry) and for a laundry circuit.

Applying these values to your own home, and using the actual nameplate values of major appliances, you can use one of several formulas to calculate your electrical load. One formula for homes with 120/240-volt service of 100 amps or more is presented as a worksheet in **Table I**.

To show how to use the formula, let's take the example of a house with 1,800 square feet (outside dimensions) of finished living space and space adaptable for future use. We will assume the house has the usual two small appliance circuits, a laundry circuit, a hot water heater (5,500 watts), a dryer (5,600 watts), a dishwasher (1,500 watts), a garbage disposal (600 watts), a range (15,000 watts), and a central air conditioner (5,000 watts).

The first step is to multiply 1,800 square feet by 3 watts per square foot. This gives us 5,400 watts

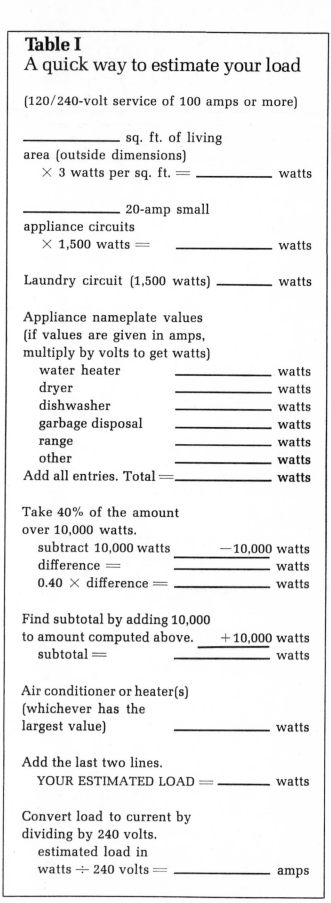

Table I
A quick way to estimate your load

(120/240-volt service of 100 amps or more)

_____ sq. ft. of living area (outside dimensions)
× 3 watts per sq. ft. = _____ watts

_____ 20-amp small appliance circuits
× 1,500 watts = _____ watts

Laundry circuit (1,500 watts) _____ watts

Appliance nameplate values (if values are given in amps, multiply by volts to get watts)
water heater _____ watts
dryer _____ watts
dishwasher _____ watts
garbage disposal _____ watts
range _____ watts
other _____ watts
Add all entries. Total = _____ watts

Take 40% of the amount over 10,000 watts.
subtract 10,000 watts _____ −10,000 watts
difference = _____ watts
0.40 × difference = _____ watts

Find subtotal by adding 10,000 to amount computed above. +10,000 watts
subtotal = _____ watts

Air conditioner or heater(s) (whichever has the largest value) _____ watts

Add the last two lines.
YOUR ESTIMATED LOAD = _____ watts

Convert load to current by dividing by 240 volts.
estimated load in watts ÷ 240 volts = _____ amps

for lighting and general purpose circuits. Then we add 3,000 watts for the small appliance circuits plus 1,500 watts for the laundry circuit. The total so far is 9,900 watts. Next we add in the nameplate values of all the major appliances except the air conditioner; we now have a total of 38,100 watts.

Our next step is to multiply 40 percent by the amount over 10,000 watts (0.40 x 28,100 = 11,240 watts). Adding the 10,000 watts to the 11,240 watts, we get a subtotal of 21,240 watts.

The final step is to add the 5,000 watts of the air conditioner. This gives us a grand total of 26,240 watts.

In terms of current, dividing this grand total of 26,240 watts by 240 volts gives us 109.33 amps. The service rating for our sample house, therefore, should be 125 amps or higher.

Now you can try it yourself: enter the values appropriate to your own home in the worksheet and compare your total load with your present service rating. If the two values are close together, you know you won't be able to add many new loads with your present service.

What if you have less than 100-amp service?

If your service rating is less than 100 amps, you can't use the formula given in **Table I** to calculate your load. You can, however, use a different formula that incorporates the same NEC values for typical usage. Therefore, the general purpose circuits, small appliance circuits, and laundry circuit are computed exactly as they are in the first three entries of **Table I**.

Once you have figured the general purpose circuit load (3 watts x number of square feet of living area), add 1,500 watts for each 20-amp small appliance circuit and laundry circuit. Using this total, take the first 3,000 watts and add 35 percent of the amount over 3,000 watts: [3,000 + 0.35 (total − 3,000)].

Add to this value the nameplate rating of all major appliances (space heater, garbage disposal, dishwasher, etc.). This gives you your estimated load in watts. You can find the current by dividing the total wattage by your voltage — 120 volts for two-wire service or 240 volts for three-wire service.

Changing present wiring

Once you know what you have to work with, you can start planning additions and changes to your electrical system. Older homes with two-wire service of less than 100 amps simply can't support many modern electrical appliances. So if you're thinking of adding a large load such as a range or dryer, you will need to increase your service to three-wire at 100 amps or more.

If you are confused by the load calculations and are unsure whether you can add a new load, prepare a list of all your major appliances and their nameplate ratings, the square footage of your home, and the type of service you have. With this list in hand, call or go to see your local inspector and explain what you want to add. The inspector can tell you quickly if your present service is sufficient or what is the best service for your needs.

WHAT CAN YOU DO?

To get more mileage out of your present electrical system, you can do one of three things. You can extend an existing circuit, add a new circuit, or add a subpanel. Remember: the number of receptacles, circuits, or subpanels is not important. What *is* important is your total house load; it must not exceed your service rating.

Extend a circuit

Perhaps the easiest way to add to a wiring system is to extend an existing circuit. This would be useful if you find yourself depending on extension cords. Extension cords are for temporary use only; they cannot be used as permanent wiring.

A circuit can be tapped wherever there is a box (such as where there is a receptacle, switch, or fixture outlet). One exception to this is a switch box that is wired with two hot wires only, as in the case of a switch loop (see drawing **13-D**).

When tapping into a circuit at a switch, you must know which hot wire is from the source. If a circuit extension is wired inadvertently into the hot wire from a switch, the switch will control the extension.

Drawing **17-A** shows how to extend a circuit to a receptacle from a switch. Note that the hot wire going to the receptacle is spliced to the hot wire from the source. Drawing **16-D** shows how to extend a circuit from a light fixture; drawing **19-A** shows how to extend a circuit from a receptacle.

When extending a circuit, make sure you maintain the integrity of the circuit wiring and its overcurrent protection device. In other words, be sure to use the same size wire for the extension as the existing circuit wires (assuming the existing circuit wires are the correct size).

Add a new circuit

Adding a new circuit is often the answer when an existing circuit can't handle a new load or when a new appliance requires its own circuit. Before you add a new circuit, though, calculate your total house load including the new load to make sure you will still be within your service rating. Note that all new 120-volt branch circuits must have a grounding wire and must comply with present code requirements (see "Branch circuit requirements").

Don't think that just because your distribution center is completely full of fuses or circuit breakers, you can't add any new circuits. If your panel uses breakers, one option is to replace one 120-volt breaker with a 120-volt, two-circuit breaker especially designed to fit in the space of one breaker (see drawing **32-A**).

Another option to consider, whether you have fuses or circuit breakers, is adding a subpanel. To do this, remove two branch circuits from your distribution center to make room for a two-pole breaker for subfeeds; then route the subfeeds to the new subpanel. There is no limit to the number of subpanels you can have, as long as your load doesn't exceed your service rating.

For a discusssion on subpanel wiring, see pages 64-66.

Planning your new wiring

Planning is a crucial first step in putting in a wiring system from scratch. Time spent at this stage can save you both time and money when you do the actual wiring.

Branch circuit requirements

Start your planning by reviewing the following National Electrical Code branch circuit requirements. Remember that these are *minimum* requirements; more circuits and outlets can (and, in many cases, should) be added.

Kitchen and related areas. The area of a home most restricted by the Code includes the kitchen, dining

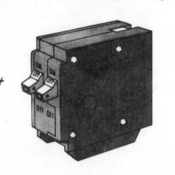

32-A

120-volt, two-circuit breaker

Can I plug in my new appliance?

Rather than blowing a fuse first, you can easily calculate whether a new appliance can be plugged into an existing circuit. Let's say you just bought an electric skillet and you want to know if you can plug it into the same circuit with a toaster and use both for breakfast. (Your circuit map told you what else is on the circuit.)

First you want to find the total amount of current drawn by the two appliances. The nameplates will tell you either the wattage or the current (amps) for the appliances. If wattages are given, take the total wattage and divide by the voltage to obtain the current. If the new skillet says 1,250 watts and the toaster says 900 watts, the total is $1,250 + 900 = 2,150$ watts.

The voltage of the circuit is 120 volts, so the amperage (current) is 2,150 watts ÷ 120 volts = 17.9 amps. If the appliances are rated in amperes, the division is already done for you; just add the numbers of amps listed.

Now that you know how much current will be drawn, check to see if that amount is within the circuit rating. To find the circuit rating, look either at your circuit map or at the fuse or breaker protecting the circuit. A kitchen circuit is probably rated for 20 amps, so with appliances having these wattages you'll be able to enjoy ham, eggs, and toast all cooked at the same time.

The principle of comparing load to rating is the same for any circuit. Find the total current drawn by all devices which may be used together (such as lights, sewing machine, small heater, etc.) by adding the values listed on the appliances. When wattages are listed, use the following formula to find the current: total wattages ÷ voltage = total amps. Then compare this value with the rating of the circuit. When the answer is greater than the circuit rating, you'll have to plug the new appliance into a receptacle on another circuit.

A restriction

If the current load you're planning to use will be switched on for more than three hours straight, the circuit rating must be reduced to 80 percent of its normal value. A normal 20-amp circuit is then rated for 0.8×20 amps = 16 amps.

This situation might arise, for instance, when using an electric heater which runs continuously. The Code makes this restriction to prevent detrimental buildup of heat in the circuit conductors and equipment.

room, family room, breakfast room, and pantry. The receptacles in this area must be served by at least two 20-amp small appliance circuits. These receptacles are intended for small appliances, such as those used in food preparation and refrigeration. No light fixtures or other outlets can be connected to these circuits.

The receptacles should be evenly distributed between the small appliance circuits. That is, if there are two small appliance circuits and eight receptacles in the kitchen, four receptacles should be on one circuit and four on the other.

If your kitchen will have a dishwasher and/or garbage disposal, you must provide a separate 20-amp circuit for each of these appliances. An electric range should be supplied by an individual 50-amp, 120/240-volt major appliance circuit.

Laundry area. The Code requires a separate 20-amp circuit to supply the receptacle for a washing machine. If your laundry equipment will also include an electric dryer, you will need an individual 30-amp, 120/240-volt major appliance circuit for it.

The rest of the house. Circuit requirements for the living room, bedrooms, and bathroom are not as specific as they are for the previously mentioned parts of a house. In fact, once you've assigned the circuits for the restricted areas, the rest of the circuits can be 15-amp general purpose circuits. These circuits are used to supply both light fixtures and receptacles. (This excludes other major appliance circuits, such as for a water heater or a central heating system.)

By dividing 500 square feet into the total square footage of your home, you can find out how many 15-amp general purpose circuits you should have. In other words, if your home is 1,500 square feet (based on outside measurements), you should allow at least three 15-amp circuits. If you have 1,600 square feet, allow at least four.

Required receptacles and outlets

Now, for a closer look at the circuits, let's discuss specific receptacle and lighting outlet requirements.

Receptacles. For most areas of a house, the required number of receptacles depends on the size of each room. Any wall space two feet or more in width must have a receptacle. Receptacles must be spaced not more than 12 feet apart and not more than 6 feet from each door or opening (this includes archways but not windows). This would allow a lamp or appliance with a 6-foot cord to be used near any wall without an extension cord.

When a receptacle is located behind a stationary appliance such as a refrigerator, it is not considered as one of those required every 12 feet.

In the kitchen and eating areas, every counter space wider than 12 inches should have a receptacle. Every basement is required to have at least one receptacle.

GFCIs. At least one receptacle is also required near the water basin in a bathroom, on the outside of the house, and in the garage. These areas require special treatment, however, because of the possibility that you might contact a grounded metal plumbing fixture or a concrete patio floor at the same time you are using a defective electrical appliance. For this reason, bathroom, outdoor, and garage receptacles must be protected by ground fault circuit interrupters (GFCIs).

The Code permits you to install GFCI protection in either of two ways. You can use a receptacle with a GFCI built right into it, or you can install a GFCI in the distribution center in place of the circuit breaker protecting that particular circuit. For more information about GFCIs, see page 41.

Lighting outlets. Required lighting outlets fall into two groups: those that must be controlled by a wall switch, and those that may have any kind of turn-on arrangement.

The Code states that every room, hallway, stairway, attached garage, and outdoor entrance must have at least one lighting outlet controlled by a wall switch. However, in rooms other than kitchens and bathrooms, the wall switch can control one or more receptacles (which lamps can be plugged into) rather than an actual lighting outlet such as a ceiling or wall-mounted light.

In the "any kind of switching" category, the Code requires one lighting outlet in the utility room, attic, basement, and underfloor space where used for storage or containing equipment that may require servicing.

NEW SERVICE CONSIDERATIONS

You can figure the service rating you'll need the same way the load calculations are done in **Table I.**
If you're uncertain of the wattages of some of the appliances you plan to have, refer to "Typical wattages at a glance," page 34.

Loads consisting of less than 10,000 watts, with no more than five two-wire circuits, can have service smaller than 100 amps, but not less than 60 amps, according to the NEC. This situation might occur in a cabin or small vacation home. For all practical purposes, however, the minimum size is 100-amp, three-wire service which can deliver 24,000 watts.

Higher service ratings are also available, depending on your electrical load. For example, the next larger standard ratings are 125, 150, and 200 amps. Don't cut any corners when estimating your new service rating. In fact, you should leave an extra margin of service for the future. It is much easier and cheaper to install larger service entrance equipment the first time than it is to increase your service a short time later.

Usually the local utility company will specify the location of your service entrance. The utility will also tell you whether the service drop will be overhead or underground.

The type of service equipment you need will depend both on its location and on how you plan to run your circuits. If your service entrance is centrally located, for example, you will probably want to run all branch circuits directly from it. On the other hand, if your service entrance panel is in an out-of-the-way spot, it may be preferable to have a smaller service entrance panel and use subpanels, fed by a set of subfeeds from the main panel, elsewhere in your home and garage. By placing subpanels at areas of high usage, such as the kitchen and the workshop, you can route your branch circuits from the subpanels rather than routing all the circuit runs from the service entrance panel. Since this method means shorter, more direct circuit runs, it saves both time and material costs.

For information on installing service equipment and subpanels, see pages 61-64.

CIRCUIT RUNS

One approach to planning your branch circuits is to draw a diagram showing the location of each proposed receptacle, light fixture, and major appliance. Refer back to the symbols shown in "Circuit mapping," page 29, to make this easier.

Once you know where you will want power, the next step is to design the various circuits. When you do this, keep in mind that it is unwise to have a single circuit supplying the lights for an entire section or floor of a house. Plan your circuits so you won't be left in the dark in the event that one circuit fails.

Plan your circuit routes behind walls, below floors, and above ceilings. Make your routes as direct as possible, following structural building members wherever you can.

One option to consider when planning branch circuits is to run a 240-volt, three-wire circuit to a junction box and branch off from there into two 120-volt, two-wire runs for receptacles and/or light fixtures (see page 66). Referred to as "split-circuit" or "multiwire" wiring, this use of a 240-volt circuit allows you to run the equivalent of two 120-volt branch circuits with less materials and time.

See pages 68-79 for information on how and where to make circuit runs using the various wiring methods.

Table II Typical wattages at a glance

Appliance	Wattage
Air conditioner, room	800-1,600
Blender	350-1,000
Broiler	1,000-1,500
Can opener	100-216
Coffee grinder	85-132
Coffee maker	850-1,625
Corn popper	600
Crockery cooker	110-250
thermostatic models	1,600
Dishwasher	1,080-1,800
Drill, portable	360
Dryer, clothes	5,600-9,000
gas dryer	660
Electric blanket (full size, single control)	200
Fan, exhaust (for range)	176
Fan, portable	100
Food processor	200
Freezer, frostless	1,056
Freezer, standard	720
Frying pan	1,250-1,465
Furnace, fuel-fired	800
Garbage disposal	300-900
Hair dryer, hand-held	260-1,200
Heater, built-in (baseboard)	1,600
Heater, portable	1,000-1,500
Heating pad	75
Heat lamp	250
Hot plate, two-burner	1,650
Lamps, fluorescent (per bulb)	15-75
Lamps, incandescent (per bulb)	25-200
Microwave oven	975-1,575
Mixer, portable	150
Mixer, stand	225
Projector, movie or slide	350-500
Radio	100
Range	8,000-15,000
Range, cooktop	4,000-8,000
Range, oven	4,000-8,000
Refrigerator, frostless	960-1,200
Refrigerator, standard	720
Roaster	1,425
Sander, portable	540
Saw, circular	1,200
Sewing machine	75-150
Shaver	12
Soldering iron	150
Steam iron	1,100
Stereo, hi-fi	
separate receiver	420
turntable	12
turntable-receiver	50-75
Sunlamp	300
Television, black and white	250
Television, color	300
Toaster	800-1,600
Trash compactor	1,250
Vacuum cleaner	250-800
Waffle iron	1,200-1,400
Washer, clothes	840
Water heater	4,000-5,000

Tools & materials: a shopping guide

One trip to an electrical supply store is all it takes to rattle most Saturday electricians—the range of materials and gadgets to choose from is enough to short-circuit the best of minds. In this chapter we'll discuss only those items you're likely to use in wiring projects around your home. Make sure that all materials you use have the stamp or label of an electrical materials testing laboratory and that they comply with regulations in your local code.

Tools

Though you can probably do most of your work using general purpose tools such as pliers, scissors, and a pocket knife, you'll work better and faster if you invest a few dollars in the following specialized tools.

Lineman's pliers are an electrician's basic tool. Serrated jaws hold wires firmly, and just behind the jaws are a set of wire cutters.

Usually, the serrated portion is used to twist the bare wires (clockwise), then the cutters are used to snip off the last ⅛ inch or so of the twisted ends to allow a wirenut to thread evenly.

Most lineman's pliers are provided with insulated handles — a nice extra safety feature, though it's no substitute for making sure the circuit you're working on is dead.

serrated jaws

wire cutters

insulated handles

Diagonal cutting pliers, or "dikes" as they are popularly known, are used for cutting #6 and smaller wires.

Wire strippers are available in several designs, ranging from a simple two-piece scissors type to a complex, multiple-action, self-clamping type.

Fish tape is a must whenever you are going to pull wires or cable behind walls or through conduit. Made from a long piece of flattened spring steel wire, 25 and 50-foot fish tapes come on a reel for easy handling.

Conduit bender is a necessity if you expect to bend conduit. You may wish to purchase one if you have a lot of bending to do. However, if you need to make only one or two bends, you can probably rent a bender.

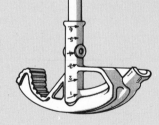

Cable ripper is a simple, low-cost tool that does a good job of ripping the outer insulation sheath on two-wire nonmetallic sheathed cable, with or without ground.

Made of a U-shaped piece of

spring steel, the ripper has a small, triangular blade in its throat which penetrates the sheath when the jaws are pressed around the cable. When you pull the tool along the cable and off the end, it leaves a slit in the insulation that allows you to peel off the outer sheath. One word of warning: use this tool only on flat, two-wire cable with or without ground.

blade

Multipurpose tool comes in many different designs and configurations. Popular designs offer as many as seven tools in one.

You should shop carefully for such a tool because various designs have strong points and weak points.

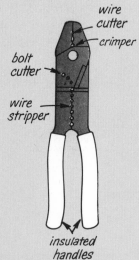

wire cutter

crimper

bolt cutter

wire stripper

insulated handles

Ordinary tools you might need:

needle-nose pliers
screwdrivers
hammer
keyhole saw
hacksaw
chisel
tape measure
electric drill and bits

Testing devices

Two basic diagnostic tools are available to help in electrical work. The first is a neon voltage tester; the second is a continuity tester.

A neon voltage tester can be used to determine whether or not a circuit is hot. Grasping the tester leads only by the insulated areas, touch one probe to a hot wire or terminal and the other to the neutral wire or terminal. If the tester lights, the circuit is hot.

You can also use this tester to determine which is the hot wire of a two-wire circuit with ground. In this case, touch one probe to the grounding wire or metal box and touch the other probe to the other wires, one at a time. The tester lights when the second probe touches the hot wire.

The tester probes can easily be inserted into the blade sockets of a duplex receptacle. When you've just unplugged a fixture or appliance because it didn't work, the tester can tell you whether the appliance is at fault or the circuit is dead.

When you are using a neon tester you are near live wires. Be very careful. Don't touch any metal parts with your hands. Use probes carefully. A carelessly placed probe can cause a short circuit if it accidentally touches both a hot and a grounded object at the same time.

Continuity tester is available in several forms. One form contains a battery and light; another uses a battery and a buzzer or bell. Use either form of continuity tester to tell whether a circuit is open or broken, or whether a short circuit exists. Before you use a continuity tester, make very sure the power is off — either turn off the breaker or pull the fuse.

How do you use this tester? Let's say you have just completed a circuit from the distribution center to a new light fixture and you wish to check out the circuit before turning on the breaker. Put a continuity tester between the breaker end of the hot wire and the ground (clip the alligator clip of the tester to the neutral bus bar and touch the end of the hot wire with the tester probe). With no light bulb in the new fixture, have someone turn the switch on and off. If the tester does not light (or buzz), you know you have not shorted the hot wire. Now put the light bulb in the fixture and have your helper repeat the switching. If the continuity tester lights (or buzzes) with the switch on and does not with the switch off, you have done a good job. You can then go ahead and put the new hot wire into the circuit breaker terminal and energize the circuit.

A continuity tester can also be used to determine whether a cartridge fuse is good or has blown.

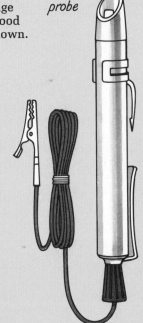

tester probe

Single conductors

A single conductor is an individual wire, usually sheathed with an insulating material. We say "usually" because a grounding wire may be bare (see below).

American Wire Gauge (AWG) numbers assigned to electrical wires indicate the diameter of the metal conductor only, not including the insulation.

Conductors are shielded from one another by material that does not carry current — color-coded thermoplastic. White or gray insulation indicates neutral wires, green is for grounding wires, and all other colors (red, black, blue, etc.) are used to identify hot wires.

Although copper is the best and most commonly used metal for conductors, aluminum and copper-clad aluminum are also used. Because aluminum is not as efficient a conductor as copper, aluminum or copper clad aluminum wire must be larger than a copper wire to conduct the same amount of electricity.

To assure a good connection when using #6 or larger *aluminum* conductors, smear an oxide inhibitor on the end of the conductor first, then tighten the terminal. Go back the next day and tighten the terminal once again.

Electrical codes take the guesswork out of conductor selection by prescribing wire use. Check your local code.

Table IV and **Table V** list the ampacity (current-carrying capacity) recommended by the National Electrical Code. These tables apply when there are no more than three current-carrying conductors in a cable or enclosure (see page 42).

CROSS SECTIONS OF COPPER CONDUCTORS

actual size

18 16 14 12 10 8 6 4 2 1/0 2/0

Low voltage for thermostats, doorbells, and small appliance wiring and cords.

120/240 volts for lighting and wall outlet circuits.

240 volts for appliance circuits, service entrance, and subfeeds.

Multi-conductor cables

Cable is identified by the size and number of conductors it contains. For example, a cable with two #14 wires (one white and one black) and a grounding wire (green or bare) is called a "14-2 with ground."

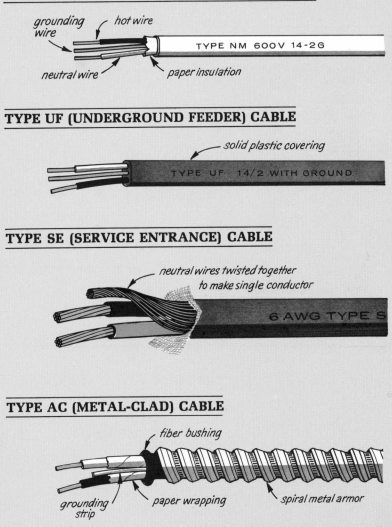

TYPE NM (NONMETALLIC SHEATHED) CABLE

grounding wire

hot wire

TYPE NM 600V 14-2G

neutral wire

paper insulation

TYPE UF (UNDERGROUND FEEDER) CABLE

solid plastic covering

TYPE UF 14/2 WITH GROUND

TYPE SE (SERVICE ENTRANCE) CABLE

neutral wires twisted together to make single conductor

6 AWG TYPE S

TYPE AC (METAL-CLAD) CABLE

fiber bushing

grounding strip

paper wrapping

spiral metal armor

Flexible cords

Flexible cord is used to connect appliances, lamps, and portable tools to outlets. *It may never be used as permanent wiring or as a permanent extension of fixed wiring.*

LAMP OR FIXTURE CORD

type SPT ("zip"cord)

molded thermoplastic insulation

POWER CORDS

type SJT

thermoplastic insulation

SJT

type SVT

vacuum cleaner cord

HEATER CORDS

type HPN

looks like heavy version of zip cord

type HPD

asbestos insulation

cotton or rayon braid

Conduit

Designed specifically to enclose and protect, conduit shields conductors from moisture and physical harm. Conduit is sized according to its inside diameter, coming in sizes ranging from ½ inch to 6 inches. The size conduit you need depends on the number and size of conductors it will be holding.

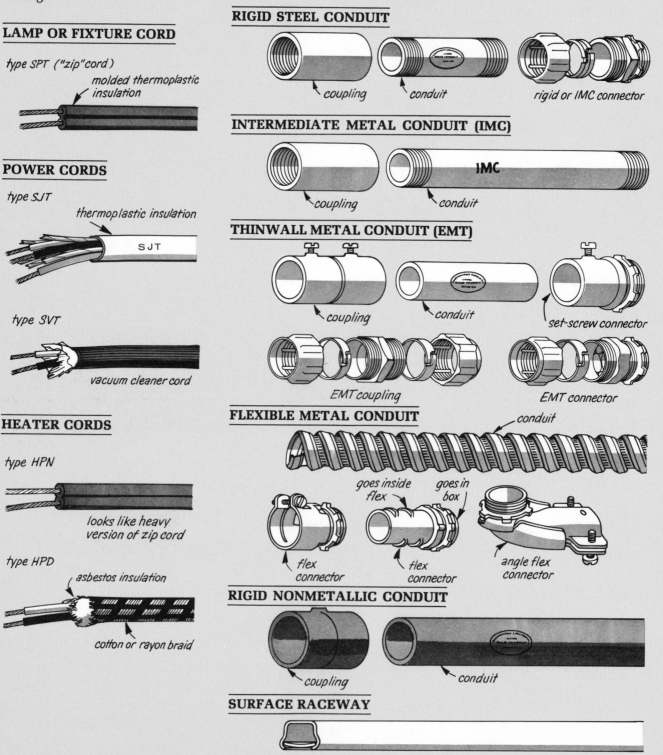

RIGID STEEL CONDUIT

coupling · conduit · rigid or IMC connector

INTERMEDIATE METAL CONDUIT (IMC)

coupling · conduit · IMC

THINWALL METAL CONDUIT (EMT)

coupling · conduit · set-screw connector

EMT coupling · EMT connector

FLEXIBLE METAL CONDUIT

conduit

goes inside flex · goes in box

flex connector · flex connector · angle flex connector

RIGID NONMETALLIC CONDUIT

coupling · conduit

SURFACE RACEWAY

Electrical boxes and accessories

Boxes are simply connection points, either for joining wires or for connection to outside devices such as receptacles, switches, and fixtures. Regardless of general trade terminology, most boxes are interchangeable in function. For example, with appropriate contents and covers, the same box could be used as an outlet box, a junction box, or a switch box. The variety of sizes and shapes corresponds to variations in wiring methods, kind and number of devices attached to the box, and number of wires entering it. One important factor is that boxes come in both metal and nonmetallic versions.

Each box has a certain volume in cubic inches that determines how many wires of a certain size may be brought into it.

Table VII, page 42, shows the number of wires you may bring into any particular size box. If the box you have chosen is too small for the number of wires needed, you can either replace the box with a deeper or larger one, or (if you can stand the appearance) you can enlarge the box with an extender ring.

EXTENDER RING

Essentially a box with the back cut out, an extender ring screws onto another box in piggyback fashion.

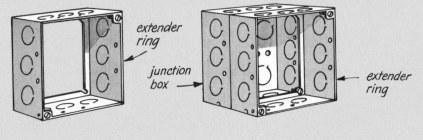

extender ring

junction box

extender ring

SWITCH BOXES

Boxes that hold only switches and receptacles are called "switch boxes." Some metal switch boxes have removable sides and can be "ganged" together to form a box large enough to hold more than one device.

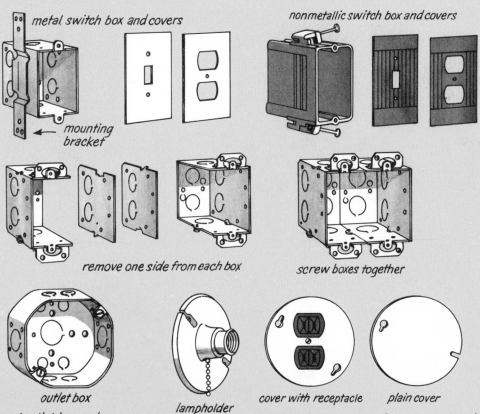

metal switch box and covers

nonmetallic switch box and covers

mounting bracket

remove one side from each box

screw boxes together

OUTLET BOXES

Outlet boxes are usually octagonal or square. They are used to hold devices, mount fixtures, and protect wire connections.

outlet box

octagonal outlet box and covers

lampholder

cover with receptacle

plain cover

(Continued on next page)

...Outlet boxes (cont'd.)

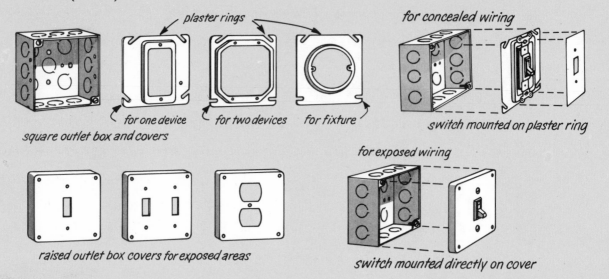

plaster rings

for one device *for two devices* *for fixture*

square outlet box and covers

for concealed wiring

switch mounted on plaster ring

raised outlet box covers for exposed areas

for exposed wiring

switch mounted directly on cover

JUNCTION BOXES

When an outlet box contains only wire splices or cable connections — no devices — it's topped with a plain cover and referred to as a "junction box."

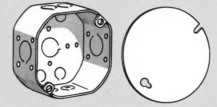

octagonal junction box and cover

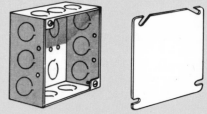

square junction box and cover

SPECIAL SITUATION BOXES

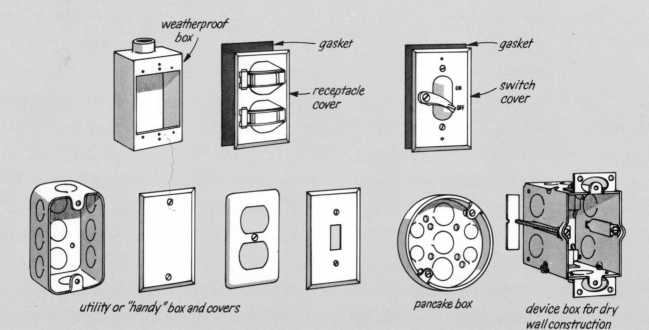

weatherproof box

gasket

receptacle cover

gasket

switch cover

utility or "handy" box and covers

pancake box

device box for dry wall construction

Switches

Some switches make a clicking sound and some are silent. Read the information on the package to assure that you get the kind you want. Note that all switches are rated according to the specific amperage and voltage they are suited for. Switches marked CO-ALR can be used with either copper or aluminum wire. Unmarked switches and those marked CU-AL may be used with copper wire only.

SINGLE-POLE SWITCH

Identified by two terminals and the words ON and OFF printed on the toggle, a single-pole switch controls a light or receptacle from one location only.

THREE-WAY SWITCH

Identified by three terminals and plain toggle, three-way switches operate in pairs to control a light or receptacle from two locations.

FOUR-WAY SWITCH

Identified by four terminals and no ON or OFF indicators on the toggle, a four-way switch is used only in combination with a pair of three-way switches to control a light or receptacle from more than two locations.

DIMMER SWITCH

This switch allows you to get maximum or minimum brightness from a light, or any gradation in between. (Note: special dimmer switches are required for fluorescent lights.)

Receptacles

The Code requires that all receptacles for 15 or 20-amp, 120-volt branch circuits (most of the circuits in your home) be of the grounding type shown below.

Like switches, all receptacles are rated for a specific amperage and voltage. Receptacles marked CO-ALR can be used with either copper or aluminum wire. Unmarked receptacles and those marked CU-AL may be used with copper wire only. Be sure you buy what you need.

To eliminate the possibility of plugging a 120-volt appliance into a 240-volt receptacle, higher-voltage circuits use special receptacles and matching attachment plugs.

DUPLEX RECEPTACLE

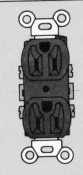

SPECIAL RECEPTACLES

air conditioner plug: 30 amps, 250 volts

dryer plug: 30 amps, 125/250 volts

range plug: 50 amps, 125/250 volts

Ground fault circuit interrupter

The ground fault circuit interrupter (GFCI, or sometimes GFI) is a device that protects you from electric shocks.

A GFCI monitors the amount of current going to and coming from a receptacle (or, in some cases, an entire circuit). Whenever the amounts of incoming and outgoing current are not equal — indicating current leakage (a "ground fault") — the GFCI opens the circuit instantly, cutting off the electricity.

The protection value of a GFCI lies in its quick response and sensitivity. GFCIs are built to trip in 1/40 of a second in the event of a ground fault of 0.005 ampere.

CIRCUIT BREAKER GFCI

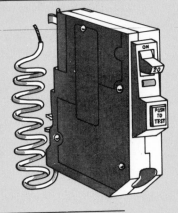

RECEPTACLE GFCI

Table III Three common thermoplastic-insulated conductors	Type	Max. operating temperature	Application
	TW	60 C, 140 F	Dry and wet locations
	THW	75 C, 167 F	Dry and wet locations
	THWN	75 C, 167 F	Dry and wet locations

Table IV Ampacity of insulated copper conductors

Wire size	Insulation type	Ampacity
14	TW, THW, THWN	15
12	TW, THW, THWN	20
10	TW, THW, THWN	30
8	TW	40
8	THW, THWN	45
6	TW	55
6	THW, THWN	65
4	TW	70
4	THW*, THWN*	85
2	TW	95
2	THW*, THWN	115
1	THW, THWN	130
2/0	THW*, THWN	175

*Exception - when used as service entrance conductor:

4	THW, THWN	100
2	THW, THWN	125
1	THW, THWN	150
2/0	THW, THWN	200

Table V Ampacity of insulated aluminum and copper-clad aluminum conductors

Wire size	Insulation type	Ampacity
12	TW, THW, THWN	15
10	TW, THW, THWN	25
8	TW	30
8	THW, THWN	40
6	TW	40
6	THW, THWN	50
4	TW	55
4	THW, THWN	65
2	TW	75
2	THW*, THWN*	90
1/0	TW	100
1/0	THW*, THWN*	120
2/0	THW, THWN	135
4/0	THW*, THWN*	180

*Exception - when used as service entrance conductor:

2	THW, THWN	100
1/0	THW, THWN	125
4/0	THW, THWN	200

Table VI Size of conduit

Size of wires	Number of wires				
	2	3	4	5	6
TW					
14	½"	½"	½"	½"	½"
12	½"	½"	½"	½"	½"
10	½"	½"	½"	½"	¾"
8	½"	¾"	¾"	1"	1"
THW					
14	½"	½"	½"	½"	½"
12	½"	½"	½"	¾"	¾"
10	½"	½"	½"	¾"	¾"
8	¾"	¾"	1"	1"	1¼"
TW and THW					
6	¾"	1"	1"	1¼"	1¼"
4	1"	1"	1¼"	1¼"	1½"
2	1"	1¼"	1¼"	1½"	2"
1/0	1¼"	1½"	2"	2"	2½"
2/0	1½"	1½"	2"	2"	2½"
4/0	2"	2"	2½"	2½"	3"

Table VII Number of conductors per box

Type of box	Size	Number of conductors*			
		#14	#12	#10	#8
Octagonal	4"x1¼"	6	5	5	4
	4"x1½"	7	6	6	5
	4"x2⅛"	10	9	8	7
Square	4"x1¼"	9	8	7	6
	4"x1½"	10	9	8	7
	4"x2⅛"	15	13	12	10
	4-11/16"x1¼"	12	11	10	8
	4-11/16"x1½"	14	13	11	9
Switch	3"x2"x2¼"	5	4	4	3
	3"x2"x2½"	6	5	5	4
	3"x2"x2¾"	7	6	5	4
	3"x2"x3½"	9	8	7	6

*Count all grounding wires in a box as one conductor.
Count each device as one conductor.
Count each wire entering and leaving box without splice as one conductor.
Pigtails are not counted at all.

Wiring know-how

This is a how-to chapter. In words and illustrations we explain the basic elements of working with wires: how to strip wires, how to join wires, and how to wire your new devices. Read on as we discuss code requirements and reveal some tricks of the trade.

Cutting and stripping cable and wires

Cable consists of insulated and bare wires bundled together and wrapped in an outer sheath of insulation. Before you connect a cable to a device or join it to another cable, you must cut open and remove the outer sheath, cut away all separation materials, and strip the insulation from the ends of the individual conductors.

REMOVING OUTER INSULATION

To lay open flat cable, such as two-wire NM (with or without ground), use a cable ripper (see drawing **43-A**) or knife. If you're using round, three-wire cable—as when you're wiring three-way switches—use a pocket knife, linoleum knife, or utility knife so you can follow the rotation of the wires without cutting into their insulation.

⚠️ Do not cut open cable while it rests on your knee or thigh. Try to find a flat board or wall surface. Also, don't cut toward your body — always cut away from it.

Once you've exposed the internal wires, cut off the outer sheath and any paper, string, or other separation materials. Now you are ready to strip the insulation off the ends of the individual wires.

STRIPPING WIRES

Be careful not to nick a wire when you are stripping it. A nicked wire will break easily, especially since the nick is usually right where you bend the wire to form a loop for a connection to a screw. If you

do nick a wire, it's best to redo the stripping.

Solid wires from #14 to #10 are easily stripped using a wire stripper. Drawing **44-A** shows how to use the stripper. After you've practiced the movement several times it will become quite easy.

To strip larger wires — #8 to #4/0 — use a pocket knife to take off the insulation as if you were sharpening a pencil (drawing **44-B**). Cut away from your body.

43-A: HOW TO RIP CABLE

Step 1. *Slide cable ripper up cable to top of box. Press handles together and pull down off end of cable. This will score the outer sheath. Bend the cable back to crack the score and then peel open the outer sheath of insulation.*

Step 2. *Using a knife or a pair of dikes or utility scissors, cut off opened outer sheath of insulation and all separation materials.*

44-A: HOW TO STRIP WIRE

adjustment screw

Step 1. *Set adjustment screw for gauge of wire.*

Step 2. *Hold wire firmly in left fist, with thumb extended toward end of wire. Place stripper on wire at angle and close with right hand, left thumb pressing against pivot of stripper.*

Step 3. *Rock stripper so right hand moves left and jaws of stripper move right, using left thumb as a pivot. Insulation will slide off.*

44-B: HOW TO STRIP LARGE WIRE

Joining wires to screw terminals

When making a wire-to-screw-terminal connection, strip about ½ to ¾ inch of insulation off the wire end. Using needle-nose pliers, form a half loop in the bare wire. Hook the wire *clockwise* around the screw terminal (see drawing **44-C**). As you tighten the screw, the loop will close. If you hook the wire backward (counterclockwise), tightening the screw will tend to open the loop. If you are using aluminum wire, wrap it ⅔ to ¾ of a turn around the screw.

44-C: WRAP WIRE CLOCKWISE

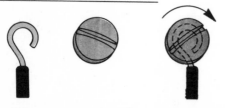

Always strip your wires so that a minimum (no more than 1/16 inch) of bare wire extends out beyond the screw head, or for that matter, beyond any connector. On the other hand, don't let the insulation extend into the clamped area.

Don't try to place more than a single wire under a screw terminal, because the terminals are made for only one wire. If you need to join several wires at a single screw terminal, use a pigtail splice (see page 45 and drawing **45-B**).

Splicing wires

Wires are joined together (spliced) with solderless, mechanical connectors. These connectors are of two basic types: wirenuts and compression rings. Note: If you must splice aluminum to copper wire, use a special two-compartment connector.

Wirenuts. These come in about four sizes to accommodate various wire combinations. Each manufacturer has its own color code to distinguish the various sizes. For example, in one brand a red wirenut can be used to splice four #12 wires or five #14 wires. Once you know how many wires of what size you'll be splicing, make sure you get the proper wirenut sizes. The packaging will tell you what wires the various sizes will take.

44-D: HOW TO PUT ON WIRENUT

Step 1. *Strip off about 1 inch of insulation from ends of wires you're going to join. Twist the stripped ends clockwise at least one and one-half turns.*

Step 2. *Snip ⅜ to ½ inch off the twisted wires so the ends are even.*

Step 3. *Screw the wirenut on clockwise.*

Compression rings. You'll need a special four-jawed crimping tool to use compression rings. Once you've stripped two or more wires and twisted them together, the tool presses a ring onto them with enough pressure to shape the metal into intimate contact. An insulating cap then fits over the connection (see drawing **45-A**).

Some jurisdictions require use of compression rings for grounding wires because they provide a more permanent bond than wirenuts. When used for grounding wires, a compression ring need not be covered with an insulating cap.

45-A: HOW TO PUT ON COMPRESSION RING

crimping tool

Step 1. *Strip off about 1 inch of insulation from ends of wires you're going to join. Twist the stripped ends clockwise at least one and one-half turns. Snip ⅜ to ½ inch off the twisted wires so the ends are even.*

Step 2. *Slip compression ring over twisted wire ends. Crimp ring with crimping tool.*

insulating cap

Step 3. *Put on insulating cap.*

compression ring

Pigtail splice. This arrangement, mentioned previously, is nothing more than three or more wires spliced together. One of the wires (the pigtail) connects to a terminal (see drawing **45-B**).

45-B: PIGTAIL SPLICE

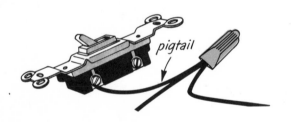
pigtail

Note: Electrician's tape should never be used in place of a wirenut or compression ring. Tape is useful for emergency insulation repairs, but it's not a substitute for a good mechanical splice.

Wiring switches

Whether you are replacing an old switch or adding new wiring, read the information stamped on your new switch carefully. Make sure the switch you are going to install has the same amp and voltage ratings as the one you are replacing or that it is suitable for the circuit. If your wiring is aluminum, be sure that the switch is designed to be used with aluminum wire (it will be identified by the letters CO-ALR).

Most switches in a home are of the single-pole or three-way types. Single-pole switches have two terminals of the same color and a definite right side up. All switches are wired into hot wires only; with a single-pole switch, it makes no difference which hot wire goes to which terminal. Because of code limitations on the number of wires that a given size box may contain, circuit wires sometimes run to the light first, with a switch loop going to the switch. This situation is illustrated in drawing **13-D.** Be sure to identify the white conductor as a hot wire by wrapping the white insulation with black tape or painting it black.

Three-way switches have two terminals of the same color (brass or silver colored) and one of another color (usually black). There is no right side up or upside down with a three-way switch; however, it is important to know which of the three terminals is the odd colored one. This terminal is often called the "common terminal."

To wire a pair of three-way switches, run the hot wire from the source to the common terminal of one switch; run the hot wire from the light to the common terminal of the other switch. Then wire the four remaining terminals by running two hot wires between the two terminals on one and the two terminals on the other. See drawings **17-B** through **18-B**.

BACKWIRED TERMINALS

Many switches and receptacles come with two sets of terminals: screw terminals and backwired terminals. If you are using screw terminals, attach the wires to the device as described on page 44 and shown in drawing **46-A.**

To backwire a device, you make the wire-to-terminal connection by simply poking a straight wire into a hole. Using the gauge that's molded into the back of the device as a guide, strip a wire and poke it in the appropriate hole (see drawing **47-A**). A jaw inside the hole allows the wire to enter, but prevents you from withdrawing it unless you release the tension by inserting a small screwdriver blade into a special slot next to the hole. Backwiring is recommended for copper and copper-clad aluminum wires only — not for aluminum wires.

When using the backwired terminals, take another minute to tighten down the unused screw terminals. This will help prevent the possibility that any loose metal will end up in the box.

⚠ De-energize circuit before working on wiring. Test to make sure circuit is dead before touching wires.

Step 1. *Install box. Secure cables to box so that 6 to 8 inches of each cable extends from box.*

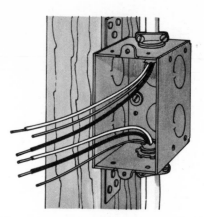

Step 2. *Strip outer sheath of insulation to back of box; remove sheath and all separation materials. Strip off ½ to ¾ inch insulation from wire ends.*

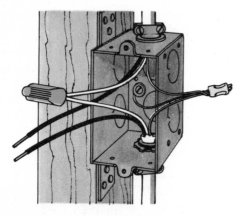

Step 3. *Join the two neutral (white) wires; cap with wirenut. Make up grounding connection by bonding the two grounding wires with compression ring or wirenut. If box is metal, add grounding jumper from box.*

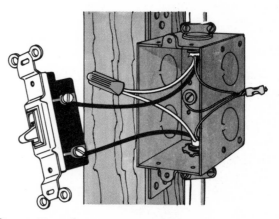

Step 4. *Form loops in ends of two hot (black) wires; wrap clockwise around terminal screws. Tighten screws.*

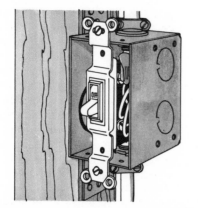

Step 5. *Push wires and switch into box. Screw switch to box.*

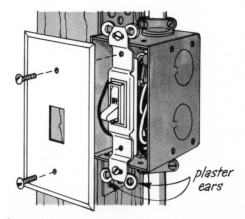

plaster ears

Step 6. *Adjust screws in mounting slots until switch is straight. If switch isn't flush with wall, remove plaster ears from mounting straps and use as shims to bring switch forward. Screw on faceplate.*

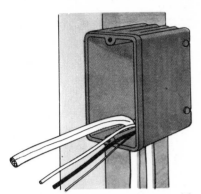

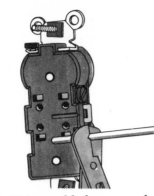

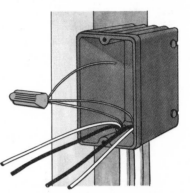

Step 1. *Install box. Insert cable coming into box from source and cable going on to next receptacle. Leaving 6 to 8 inches of each cable extending from box, secure them. Remove outer insulation sheath and separation materials from cable portions within box.*

Step 2. *Using molded gauge on back of device, measure and strip insulation off ends of wires.*

Step 3. *Make grounding connection by bonding the two grounding wires and a grounding jumper from the device with compression ring or wirenut. If box is metal, include grounding jumper from box. Remember: you must be able to pull out and disconnect a receptacle without interrupting the grounding continuity of the circuit. Also, grounding jumpers must be the same size as circuit wires.*

Step 4. *Insert wires into proper holes: neutral (white) wire in holes identified as white and hot (black) wires in opposite holes. Tighten unused screw terminals. Note: backwiring is suitable for copper wire only.*

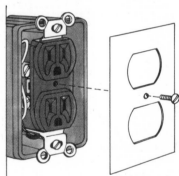

Step 5. *Push wires and receptacle into box. Screw device to box. Screw on faceplate. Wire next receptacle in circuit the same way.*

Wiring grounding type receptacles

All grounding type receptacles must be grounded, whether they are installed as replacements in existing wiring or as new work. In new work, simply connect the receptacle to the circuit wires as shown in drawing **47-A**. If, however, you're using a grounding type receptacle to replace an ungrounded receptacle in an existing circuit that doesn't have an equipment grounding wire, you must ground the receptacle independently to a cold-water pipe.

The National Electrical Code has an exception: if it's impractical to ground to a cold-water pipe, replace receptacle with a nongrounding type.

Receptacles have three different colors of screw terminals. The brass colored screws are hot terminals, the white or silver colored screws are neutral terminals, and the green screw is the grounding terminal. If the receptacle is a backwired type, it will have some kind of identification for the hot and neutral terminals. The grounding wire must still be attached to the green screw terminal.

If your wiring is aluminum, be sure that your receptacle is designed to be used with aluminum wire (it will be identified by the letters CO-ALR). Use the screw terminals only; backwiring is not suitable for aluminum wires.

A usual circuit arrangement is for several receptacles to be wired in parallel. Drawing **47-A** shows the wiring of one receptacle from such a circuit.

Occasionally you might wish to have the two outlets of a duplex receptacle operate independently of each other. For example, you might want one outlet to be controlled by a switch and the other one to be always hot. Or you may wish to wire the two outlets of a receptacle into different circuits. Using pliers, remove the break-off fin that connects the receptacle's two hot terminals. Drawing **19-C** shows a receptacle wired so that one half is controlled by a switch and the other half is always hot.

Wiring a ground fault circuit interrupter

Two types of GFCIs are used in homes: the circuit breaker type and the receptacle type. The circuit breaker type, which takes the place of a regular breaker, protects all receptacles hooked to that branch circuit (see drawing **48-A**).

A receptacle GFCI is wired like an ordinary receptacle. Any receptacle type GFCI provides protection at that particular receptacle. Depending on the model, it may also protect all other receptacles downstream (away from the source) from it, but it will not protect any receptacles upstream (toward the source) from it.

48-A: HOW TO WIRE GFCI

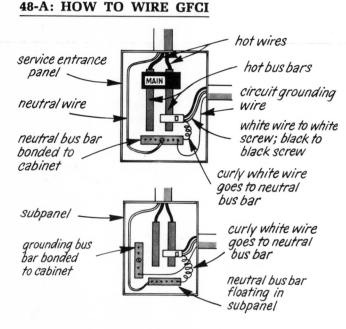

Wiring light fixtures

Light fixtures come prewired; using wirenuts, you'll find it a simple matter to join the black fixture wire to the black circuit wire and the white fixture wire to the white circuit wire. When you mount a metal light fixture on a grounded box (metal or nonmetallic), the screws holding the fixture to the

box ground the fixture — except in the case of a chain-hung fixture. When installing a chain-hung fixture, run a separate #18 grounding wire from the box to the fixture as shown in drawing **48-B**.

48-B: HOW TO WIRE CHAIN-HUNG FIXTURE

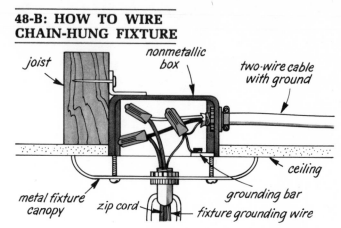

Most ceiling and wall light fixtures can be attached directly to an outlet box, much as a cover is screwed on any box (see drawing **48-C**). Some fixtures, though, are supported by a special adapter or fixture stud that mounts directly on the box (see drawing **48-D**). Adapters and fixture studs are usually supplied with the light fixture.

48-C: CEILING-MOUNTED LIGHT

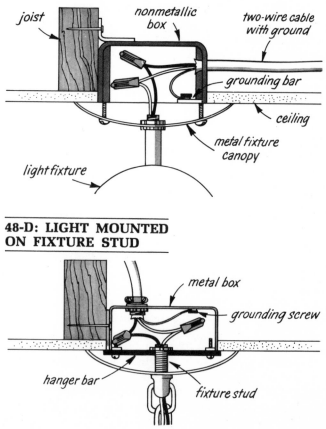

48-D: LIGHT MOUNTED ON FIXTURE STUD

Upgrading your existing wiring

Do you have extension cords trailing across rooms for plug-in appliances and lamps? Or are you fed up with grappling in the dark for a pull chain to turn on a light? If your wiring is inadequate or—worse yet—unsafe, perhaps you should consider upgrading it. This chapter offers specific techniques for installing concealed wiring in homes where wall, ceiling, and floor coverings are already in place.

Charting your course

Before you begin any work, read through this chapter to get a feel for the project ahead. Then check with your local building department about getting an electrical permit. Obtaining a permit and requesting inspection add up to a solid guarantee that the work you do is performed properly. And it's an inexpensive way to get expert advice from a pro—your electrical inspector.

SURFACE OR CONCEALED WIRING?

You can supplement your existing wiring using one of two methods: surface wiring or concealed wiring. Easier to install, but less common, is surface wiring.

Where routing wire through walls and cutting open walls, ceilings, and floors is too difficult, surface wiring is the best answer. Safe and neat, surface wiring systems usually consist of protective channels or strips that allow you to mount wiring and boxes on practically any surface (see page 38 for an illustration of surface raceway). Because surface wiring materials from various manufacturers differ somewhat, consult your electrical supplies dealer for more information about the various systems available.

The more common method of rewiring a home is concealed wiring. For this method you run cable behind walls, above ceilings, and under floors. Then you insert boxes in the walls and ceilings for your new switches, receptacles, and light fixtures.

Concealed wiring is what this chapter is about. Our discussion applies only to wood frame homes; if your house has masonry walls or floors in places where you want to wire, consider surface wiring.

CIRCUITS AND SOURCES

First of all, you should decide what you want and where you want it. Put your plans on paper; page 29 shows how to diagram your circuits.

Give some thought to possible power sources for your new wiring. You'll have to choose between adding new circuits to your distribution center or tapping into existing circuits at receptacle, light, or switch boxes. For this decision you must first evaluate your present wiring (see pages 28–31).

What are the requirements for tapping into an existing circuit? One key element is the presence of both a hot and a neutral wire that are in direct connection with the source at the distribution center. This means that you can't tap into a switch or light fixture at the end of a circuit (see page 31).

Here are two more requirements: the wires must meet in a box, and the box must be large enough to accommodate the additional three wires of the new cable (see **Table VII**, page 42). The box must also have a knockout through which you can insert the new cable.

Access is very important, too. If you have a good, accessible source, but the box isn't right, you can change the box.

Going from a two to a three-hole receptacle

Many portable appliances and tools have a grounding wire to eliminate the possibility of electric shock. A three-prong plug indicates the presence of a grounding wire. This wire connects to the circuit grounding wire through the third hole of a grounding type receptacle (see page 41). What do you do with a three-prong appliance plug when your receptacles are the two-prong variety? You can do one of two things: use an adapter plug or change the receptacle.

An adapter plug is effective only if the receptacle box is grounded and if you connect the pigtail properly. If your circuit wiring runs in conduit or is armored cable or nonmetallic sheathed cable with ground, chances are good that your box is grounded. Don't assume, however; test to be sure. After removing the receptacle faceplate, place one probe of a neon voltage tester on a hot terminal screw and the other probe on a mounting screw. If you get a strong light in the tester, the box is grounded

and you can use an adapter plug (see below).

If your box isn't grounded, you must replace the receptacle with a grounding type and run a grounding wire to the closest cold-water pipe. Use #14 copper wire for a 15-amp circuit and #12 copper wire for a 20-amp circuit.

Grounding a receptacle independently isn't always the most convenient thing to do. However, here are two suggested routes to the cold-water pipe.
1) Run wire out from box through a partially opened knockout. Then fish wire behind wall and molding to nearest cold-water pipe. Use a grounding clamp to attach the grounding wire to the pipe.
2) Make a notch in the wall covering that's covered by the faceplate. Run the wire out under the faceplate, down the surface of the wall, and along the molding to the nearest cold-water pipe. Staple the wire to the wall and molding surfaces. Attach the grounding wire to the pipe with a grounding clamp.

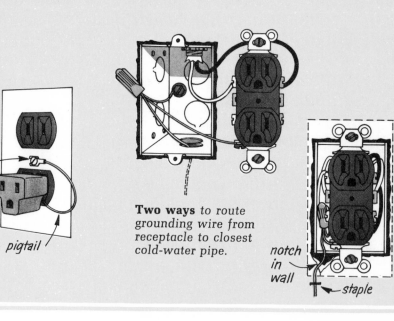

Use adapter *only if box is grounded. Plug adapter into receptacle, then screw lug on end of pigtail under faceplate mounting screw.*

faceplate mounting screw

adapter

pigtail

Two ways *to route grounding wire from receptacle to closest cold-water pipe.*

notch in wall

staple

ACCESS FOR CABLE

The next step is to study possible access routes to the proposed additions on your wiring diagram.

Wood frame homes are not all built the same way, but most have 2 by 4 stud walls, 2 by 8 (or larger) floor joists, and 2 by 6 (or larger) ceiling joists. These wooden structural members are normally spaced 16 inches apart from center to center. In some new homes, however, the spacing is 24 inches, and in some roughly-built older homes it's

somewhat random. Drawing **51-A** shows the skeleton of a typical house.

In new construction all basic wiring is done before wall, ceiling, and floor coverings are added. Rewiring a finished house, however, is a different story. You have to find ways to route cable behind walls, above ceilings, and under floors.

Explore your home's construction where you'll be adding switches, receptacles, and lights and check the routes from the power sources to these areas. Find out what you'll be dealing with. The best route for a cable is one that is direct and accessible. Accessibility is generally more important than directness. The savings in time and effort from avoiding extensive cutting and patching of walls, ceilings, and floors nearly always offsets the added material costs for an indirect cable run.

Where you have access. In some parts of your home, installing cable and boxes might be quite easy. These are areas such as attic floors and unfinished basement ceilings where wall, ceiling, and floor coverings are attached only to one side of the framing. You simply work from the uncovered side, drilling holes and threading cable through studs or joists. You can also "fish" cable through finished walls from these locations (see page 54).

Where access is limited. Getting cables into walls, floors, or ceilings that have coverings on both sides involves cutting through the coverings, installing cable, and patching. The amount and difficulty of cutting and patching depends only partly on where the cable goes; it's also determined by the surface material.

The most common wall and ceiling covering, gypsum wallboard, is relatively easy to cut away and replace. But some other materials, such as ceramic tile, some types of wood flooring and plaster, are more difficult to cut and patch and should be left alone when possible.

⚠ When in an attic, don't put any weight on the ceiling material between joists. Step only on the joists, or put planks across them for more ease. Be sure to walk gently so you don't crack the ceiling surface.

51-A: WOOD FRAME HOUSE

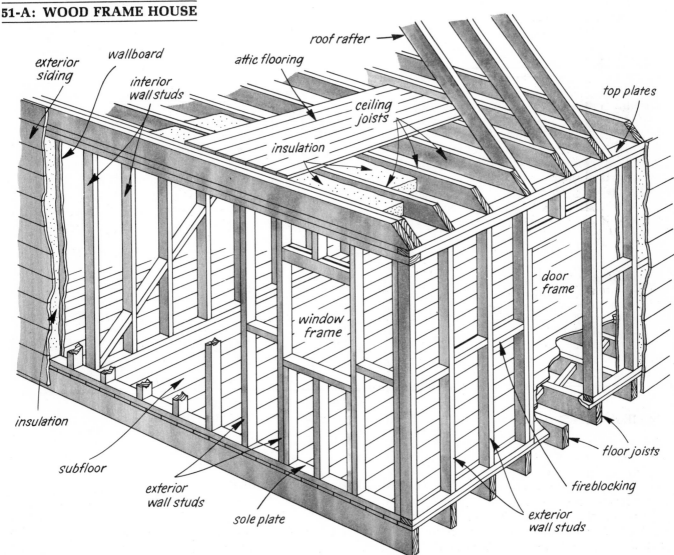

Preparing for boxes and cable

Before actually routing the cable, you must choose and buy the right boxes and cut holes in the wall or ceiling coverings to install them.

Electrical boxes for rewiring work come in many types and sizes. These boxes, sometimes called "old-work" or "cut-in" boxes, differ from many new-work boxes in that they mount easily where wall and ceiling coverings are already in place. Select a size large enough to hold all the necessary wires and the number of switches, receptacles, or combinations you want. For help in choosing sizes, consult **Table VII** on page 42.

The choice of metal or nonmetallic boxes is up to you. Metal is sturdier, but you must ground metal boxes. Nonmetallic boxes cost less and don't require grounding.

Wall boxes. For wiring in previously covered walls, you'll be choosing between cut-in boxes and plain boxes with adjustable top and bottom ears. The right type depends on the wall it goes into. Several typical styles are shown in drawing **52-A**.

Ceiling boxes. Beyond merely providing a place for joining wires and connecting them to a device, a ceiling box often supports a heavy fixture. Because of this, a ceiling box is usually fastened to a joist by a flange attached to the box or by an adjustable or an offset hanger bar.

For hanging a lightweight fixture (24 ounces or less) or a fixture with a canopy lip of ¼ inch or more, consider a ceiling cut-in box or a pancake box. All these ceiling boxes are shown in drawing **53-C**.

WHERE TO LOCATE BOXES

Plan to keep all box heights consistent. If you're putting new ones in a room that already has boxes, consider putting the new boxes at the same height as the old ones. Otherwise, place new receptacle boxes 12 to 18 inches above the floor. New switch boxes are best installed 44 to 48 inches high. The location of light fixtures is up to you — be sure they'll shed light efficiently where needed.

All boxes must be accessible — even junction boxes that contain no devices. They must not be covered by walls, ceilings, or floors.

Before proceeding, probe into the wall to find out if it's clear of wires and pipes and to be sure your box will fit in that exact location.

Check for obstructions. Hunting for studs can be an uncertain job. Sometimes you can measure to find them on 16-inch centers, but not always. You can knock on the wall with your knuckles, listening for a solid rap that indicates something behind the wall covering, but knocks often sound alike all over the wall. You might try a stud finder. These small gadgets have a magnetic pointer that's attracted to nail heads, but they work only on wooden walls or those covered with gypsum wallboard.

The surest, most accurate way to find out what's behind a wall without cutting into it is to drill a small test hole where you want the box. Bend an 8 to 9-inch length of stiff wire, push it through the hole, and revolve it (see drawing **53-A**). If it bumps something to the left or right, move over a few inches and try again until you find an empty space.

When locating a box on a plaster-and-wood-lath wall, chip away enough plaster around the test hole to expose a full width of lath. Plan to center the box on the lath.

52-A: WALL BOXES

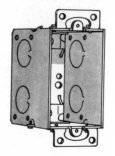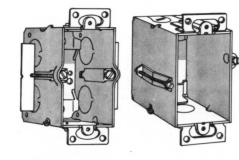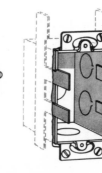

A plain box *with adjustable ears is used in wooden and plaster-and-wood-lath walls. When box is screwed directly on wooden wall, faceplate hides the ears.*

Two styles *of boxes with side clamps to be used in ⅜-inch or thinner wooden or hardboard walls.*

Combined with a pair *of brackets, a plain box is used in gypsum wallboard and plaster-and-metal-lath walls.*

Nonmetallic cut-in box *with metal spring ears is designed to be used in gypsum wallboard walls.*

53-A: PROBING FOR OBSTRUCTIONS

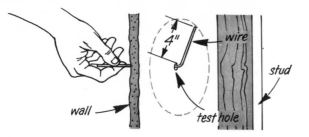

Mark hole for box. Once you've found a clear space, you can mark the wall or ceiling for cutting the box hole.

For a plain wall-mounted box, place box face down on a sheet of thick paper or cardboard to be used as a template. Trace the box's outline, omitting the adjustable top and bottom ears. For a cut-in box with side clamps, trace around the side clamps.

53-B: MARKING WALL BOX HOLE

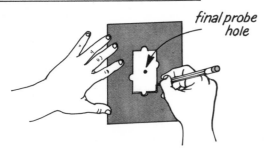

Cut the template out in the shape of the box and position it on the wall at the chosen location, traced side facing the wall. Scribe the outline on the wall (see drawing **53-B**).

A cut-in box for gypsum wallboard fits into a simple rectangular hole — just hold the box against the wall and mark around its perimeter.

For a ceiling box, if you have access above from an attic, mark the hole from above. To do this, first locate the box from below and drill a small guide hole through the ceiling. If your attic has a floor, you'll need an 18-inch extension bit for drilling up through the flooring (in this case the guide hole must be considerably larger). Next, remove any attic flooring at that spot and outline the box on the back side of the ceiling (see drawing **53-D**).

53-D: MARKING CEILING BOX HOLE

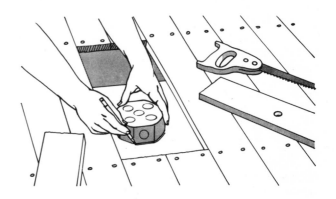

HOW TO CUT HOLES

After you've marked the holes, the next step is to cut them. Here are methods and hints for cutting the most common wall and ceiling coverings.

Gypsum wallboard. For a tidy cut in gypsum wallboard, use a saber saw or a keyhole saw.

53-C: CEILING BOXES

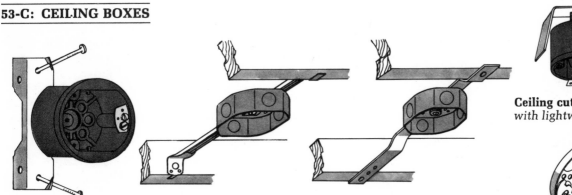

Ceiling cut-in box *is used with lightweight fixtures.*

Flange box *nails directly onto ceiling joist. This box is a good choice where you have access from above ceiling.*

Adjustable hanger bar *attaches to two joists, supporting ceiling fixture box. This box is good where you have access from above ceiling.*

Offset hanger bar *fastens to bottom edges of two joists. This box works well where you don't have access from above.*

Pancake box *is flat enough to attach directly to hanger bar or to joist. Pancake box can accommodate only five wires.*

Plaster and wood lath. Though cutting away large areas of plaster isn't recommended because of the difficulty of patching up, cutting small holes for boxes is possible.

At the hole's location, use a cold chisel to chip away enough plaster to expose one width of lath. To prevent excess plaster from cracking, tape the outside border of the hole outline with wide masking tape. Score the outline several times with a utility knife. Then drill holes as shown in drawing **54-A** for starting a keyhole or saber saw blade and turning corners. Cut slowly and evenly in the direction of the arrows (lath is usually extremely tough and fibrous; it's difficult to cut). Once you've cleared the opening, remove the tape.

When cutting a ceiling hole, brace the ceiling as you cut so large chunks of plaster don't break away.

54-A: CUTTING HOLES IN PLASTER

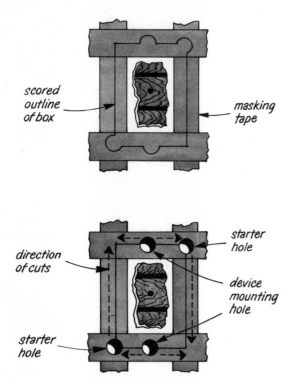

scored outline of box

masking tape

direction of cuts

starter hole

device mounting hole

starter hole

Plaster and metal lath. Trace the box outline on the wall and then tape the border with wide masking tape to prevent excess plaster from cracking. Use a metal bit to drill starter holes, then chisel away plaster within the box outline. Cut out the metal lath with a metal-cutting blade on a saber saw or a pair of dikes.

Wood. Clean cuts are easy to make in wood. Trace the box outline on the wall, drill starter holes, and then cut with a keyhole or saber saw.

Routing cable

After cutting holes but before mounting the boxes, you must run cable from the power source to the new box locations. Staple one end of the cable next to the source, leaving about a foot more than you think you'll need for the connection, then route the cable. Wait until you have all the new boxes wired and mounted before you make the actual hookup to the source.

If you're installing a new receptacle back-to-back to an existing one, routing cable is no problem. But under any other condition, this step will involve some patient work. Your goal here, as with routing cable in new construction, is to staple cable to wall studs, ceiling joists, and floor joists where it parallels them and to pass it through holes where it runs at an angle or perpendicular to them (see drawing **69-A**).

Where you have access from below or above (for example, in a basement with an unfinished ceiling or an attic with no floor), it's easy to run cable along joists or through holes in them. From these locations you can "fish" cable through walls that are covered on both sides. From a basement you can generally fish cable up to a receptacle with little trouble. Working from an attic, running cable from switches to light fixtures is usually a simple job.

HOW TO FISH CABLE WHERE YOU HAVE ACCESS

First, you'll need some fishing gear. Fish tapes (see page 35) are the tools to use for long cable runs. For shorter distances, you can use straightened coat hangers or lengths of #12 wire with one end bent into a tight, blunt hook. (Be sure that whatever you use is long enough to span the entire distance plus 2 feet.)

Next, you'll need an open space to fish through. Drawing **55-A** shows how to fish a short run where you have access from above or below.

If you're going to attempt fishing more than a foot or so, you'll need a partner. Start by following steps 1 and 2 in drawing **55-A**. Then, to make sure your passage is clear, have someone shine a flashlight in the box hole while you peer into the drilled hole to see if the light beam is visible. If it isn't, a fireblock (or something else) is in the way. For drilling through this obstruction, you'll need several extension shafts for the drill bit. Drill through the block, then look for the light again. If you still can't see the light beam, move to another location or cut away some of the wall covering and notch the block.

Once the passage is clear, follow the steps in drawing **55-B**.

55-A: FISHING SHORT RUN WITH ACCESS

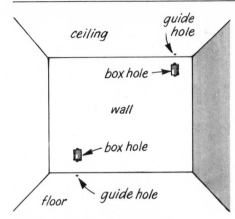

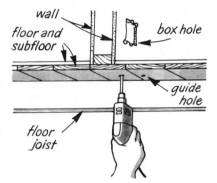

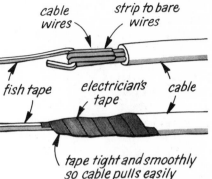

Step 1. After making a hole for your box in open space between studs, drill a small guide hole down through the floor or up through the ceiling to mark this location.

Step 2. Using a ¾-inch spade bit, drill next to your guide hole up through the sole plate from the basement or down through the top plates from the attic. Drill until you hit open space; you may have to use an extension bit.

Step 3. Run fish wire (or fish tape) in through box hole and down (or up) through drilled hole. Attach cable as shown and draw into box hole.

55-B: FISHING LONG RUN WITH ACCESS

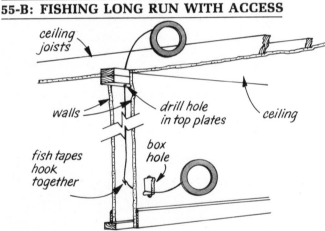

Step 1. One person runs fish tape up from box hole while the other runs a second fish tape or a length of small chain down through hole in top plates. Hook two together inside wall.

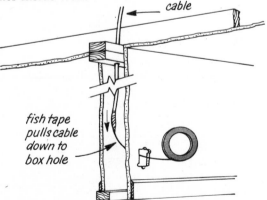

Step 2. Pull fish tapes up through wall until both partners have a secure grip on tape coming up from box hole. Attach cable to fish tape from above and, pulling slowly, work it through wall to box opening.

ROUTING CABLE WHERE YOU DON'T HAVE ACCESS

If you don't have access from an attic or basement, here are some other possible ways of routing cable.

55-C: ROUTING CABLE BEHIND BASEBOARD

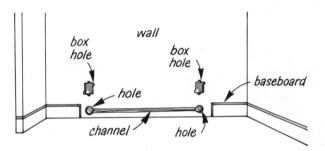

Step 1. Cut box hole or holes. Remove baseboard between box locations. Drill hole through wall below each box and cut channel in wall to connect holes.

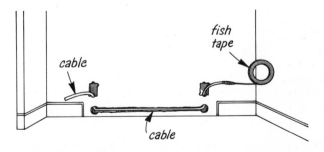

Step 2. Fish cable down through one box hole, along channel, and up into other box hole.

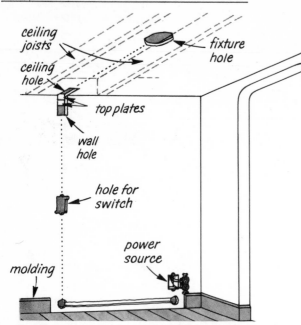

Step 1. Mark fixture box location between two ceiling joists. Cut hole for box. At ceiling edge between the same two joists, make hole in ceiling and wall where they meet. Cut hole for switch box. (Unless you want to run cable through notched wall studs or behind molding to another switch location, plan to locate switch directly below ceiling hole.) Run cable from power source to switch box.

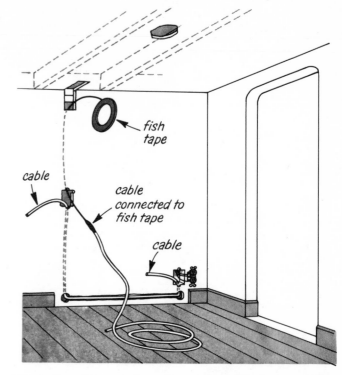

Step 2. Run fish tape down from top wall hole to switch box hole. Connect enough cable for switch loop and pull it up and out top hole.

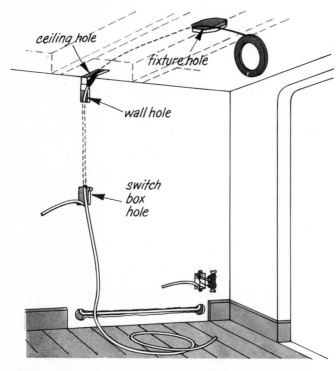

Step 3. Run fish tape from fixture hole to ceiling hole. Connect to switch loop cable and pull cable out fixture hole.

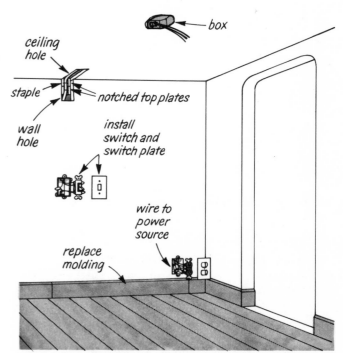

Step 4. Notch top plates to inset cable and staple cable in place. Connect cables to boxes, mount boxes, and hook up to fixture and switch.

57-A: ROUTING CABLE ALONG WALL

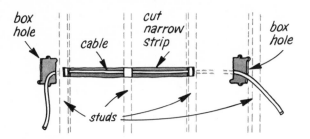

Step 1. Cut box holes in wall. Neatly cut away a straight, narrow strip of wall covering to expose all studs between box holes. Both ends of this slot should be centered over studs.

Step 2. Use a ¾-inch spade bit to drill through center of each stud. Run cable through holes.

57-B: ROUTING CABLE AROUND DOORWAY

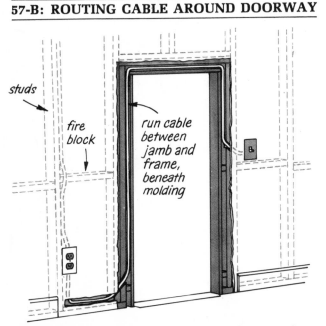

Step 1. Remove molding around door frame and as much baseboard as necessary on either side of door.

Step 2. Run cable between jamb and frame, notching spacers wherever necessary.

Working with molding. When routing cable behind molding, keep these points in mind:
• Molding may split, so be sure you can get replacements.
• Use a 4-inch-wide (or wider) putty knife or electrician's chisel to pry molding from the wall.
• Cable installed less than 1½ inches from a finished surface must be protected by a 1/16-inch metal plate or run in thinwall conduit.
• When nailing molding back up, be careful not to nail through cable.

57-C: BACK-TO-BACK DEVICES

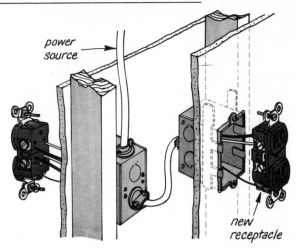

Step 1. Make hole in wall for new box. De-energize circuit you'll be working on.

Step 2. Pull device that's to be power source out of its box. Remove knockout from back of source box.

Step 3. Insert cable with connector through new box hole into source box. Connect cable to new box and mount box.

Step 4. Wire in new device. Wire into source. Turn circuit back on.

Finishing up

When the holes are cut and the cable is routed from the source to the new box locations, the only remaining jobs are mounting the boxes, making the wiring connections, and doing the patch-up work.

HOW TO MOUNT BOXES

Slip a cable connector on the cable, then insert it through a knockout in the box. Fasten the connector to the box, leaving 6 to 8 inches of cable sticking out of the box for connections. For information on stripping and preparing the wires in the box, see pages 43–44.

How you mount the box will depend on its type.

Plain box with brackets. Check the box for proper fit in the hole. If necessary, adjust the ears so the front edge of the box will be flush with the finished wall.

Put the two brackets in the wall, one on either side of the box, and pull the bracket tabs toward you so they're snug against the backside of the wall (see drawing **52-A**). Then bend the tabs over the sides of the box and secure them with needle-nose pliers.

Plain box. Check the box for proper fit in the hole. If necessary, adjust the ears so the front edge of the box will be flush with the finished wall. Mark screw placements on the lath at the top and bottom of hole. Remove box and drill pilot holes for screws. Screw box to lath (see drawing **58-A**).

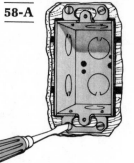

58-A

On wooden walls, just screw the ears to the wall surface. The faceplate will hide the ears and screws.

Cut-in box. Mounting this box is a one-time proposition. Once inside the wall or ceiling, the side teeth flare away from the box, making it difficult to remove. Tightening the screw at the back of the box simply pushes the teeth into the backside of the wall surface (see drawing **58-B**). Because you can't remove the box, be sure the cables are in place and the box fits the hole before you mount it. (To try out fit, remove metal spring ears from box.)

58-B

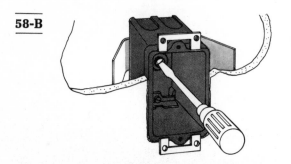

Ceiling box on hanger bar. If you don't have access from an attic, you must cut out the ceiling material between two joists. Screw the hanger bar to the sides of the two joists.

Ceiling box with offset hanger. This box works well where you don't have access from above. Screw the offset hanger bar to the bottom edges of two joists (see drawing **58-C**).

58-C

Ceiling box with flange. If you don't have access from above, you'll have to cut out a rectangle from the ceiling material. Nail or screw the flange to a joist (see drawing **58-D**).

58-D

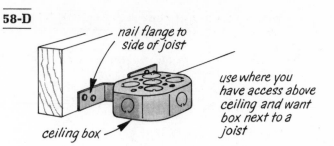

nail flange to side of joist

ceiling box

use where you have access above ceiling and want box next to a joist

Pancake box. Simply screw this box to a joist. Position the box so it will hide the hole that was drilled for the cable. You can also hang this box from a hanger bar, as shown in drawing **58-E**.

58-E

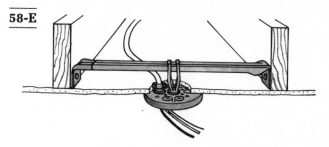

WIRING INTO THE POWER SOURCE

If your rewiring involves adding a new circuit to the service entrance panel or a subpanel, see pages 64–66.

⚠ Before hooking up your new cable to the power source, shut off the main breaker or disconnect the circuit you're wiring into.

Drawings **59-A**, **60-A**, **60-B**, and **60-C** show how to wire in a new cable at a receptacle, switch, light, and junction box respectively.

PATCHING THE HOLES

If you've notched into walls or ceilings in the process of wiring, now is the time to make them look like new. Here are some tips for putting everything back together.

Plaster walls and ceilings. The National Electrical Code requires that you repair plaster around boxes so there are no gaps or open spaces at the edge of the box. Patching around a new outlet box is a simple matter. Use a wide-edge putty knife to apply commercial plaster compound. Try to match the texture of the surrounding wall.

For larger holes you'll have to provide some backing (such as lath), clean and moisten the edges of the hole, and, in some cases, apply more than one coat.

Gypsum wallboard. For small repairs, just use a broad-bladed putty knife and some spackling compound.

To replace a larger section, you'll need to cut a new piece of wallboard to fill the hole. If there are no structural members on which to nail the replacement piece, add some wooden blocks for support.

Nail up the replacement piece, dimpling the surface slightly at the nail heads. Use a broad-bladed putty knife to spread joint compound across the dimples. With joint tape and compound, cover the edge joints around the replacement. Spread a layer of compound over the tape, being careful not to let the putty knife dip into the joint. Let the compound dry.

Apply a second coat of compound to the nail heads and tape, feathering the edges of the first coat to produce a relatively smooth surface. Sand nail dimples and joints when dry.

For a smooth wall, you may have to apply a third coat to both joints and nail heads and sand again.

To duplicate a skip-trowel texture, apply a large amount of joint compound with a broad palette knife and draw the blade over the surface in one direction. A plaster texture is applied with the same tool, moving it in a semicircular motion. Duplicate a stipple finish with a paintbrush.

Let dry, then paint the surface.

Check list for adding a receptacle

1) De-energize circuit you'll be tapping into.
2) Make hole in wall for new box.
3) Pull device that's to be power source out of its box.
4) Remove knockout in source box that leads most directly to new cable route.
5) Remove baseboard between box locations.
6) Drill hole through wall below each box.
7) Cut channel in wall between two holes.
8) Cut length of cable, adding 2 feet extra for box and device connections.
9) Put cable connector on one end of cable and fish cable through hole in wall and knockout up into source box.
10) From new box hole, fish other end of cable up into box.
11) Connect cable to new box and mount box.
12) Wire in new device. Wire into source.
13) Put faceplates on boxes and turn circuit back on.

59-A: HOW TO WIRE INTO A RECEPTACLE

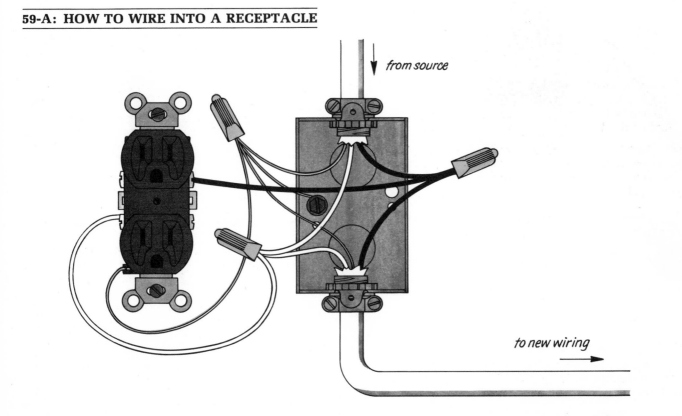

from source

to new wiring

60-A: HOW TO WIRE INTO A SWITCH

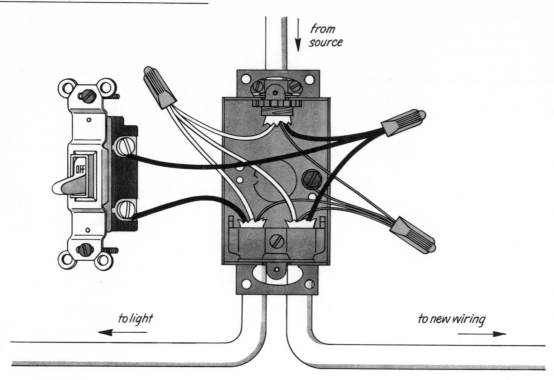

from
source

to light

to new wiring

60-B: HOW TO WIRE INTO A FIXTURE

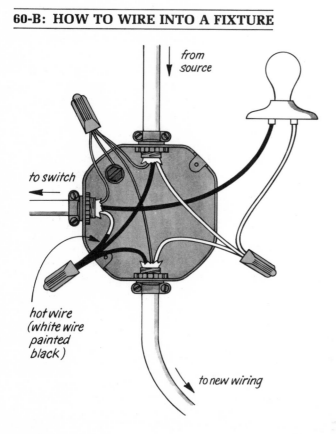

from
source

to switch

hot wire
(white wire
painted
black)

to new wiring

60-C: HOW TO WIRE INTO A JUNCTION BOX

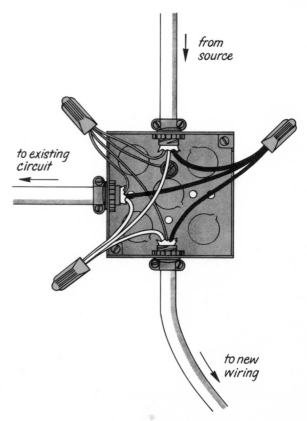

from
source

to existing
circuit

to new
wiring

Installation... step-by-step

Previous chapters have helped you make plans, select tools and supplies, and maybe even gain a little experience doing repairs. What's next? Installation. In this chapter we take you from soup to nuts—or, more precisely, from service entrance equipment to switch boxes.

Installing service entrance equipment

Your first installation considerations should be the location of your service entrance panel and whether you'll have overhead or underground service. Discuss these matters with your local utility company. Usually, the utility company specifies both the general location of your service entrance panel and such details as mast height (if overhead service) or conduit size (if underground service).

The actual sizes of service equipment and conductors, though, depend on the amount of power you'll need (your service rating) and local code requirements. So be sure to coordinate with both your utility company customer service representative and your electrical inspector before you purchase any service entrance materials.

Most service entrance panels installed today are the circuit breaker type. Therefore, our illustrations show this type of panel. The basic principles are the same for fuse type installations. In many service entrance panels, the main breaker and conductors to the meter socket are assembled at the factory.

Remember that you must completely wire your service entrance equipment (service entrance riser, main disconnect, grounding, and possibly branch circuits) and have it inspected and approved by your electrical inspector before the utility company will connect to it.

GROUNDING YOUR ELECTRICAL SYSTEM

As we have discussed before, your entire electrical system is grounded. This grounding is done at the service entrance panel (and only there) by the grounding electrode conductor and the bonding screw (see drawing **65-A**).

Grounding electrode conductor. Connect one end of the grounding electrode conductor to the neutral bus bar and the other end to an approved grounding clamp, which must be fastened securely to a metal cold-water pipe. (Note: If you have nonmetallic water pipes underground, the cold-water pipe is not a suitable ground. Check with your local inspector for the proper grounding method in this case.)

For the grounding electrode conductor, #6 bare copper wire is a good choice for up to 150-amp service; #4 bare copper is good for 200-amp service. This wire can be run through the very small knockout hole provided in the cabinet. If not concealed, it must remain next to the house until it fastens to the grounding clamp. If the conductor is subject to physical damage, it must be protected with conduit or armor.

Some jurisdictions require more than the cold-water pipe ground described here. Check with your electrical inspector about what is necessary in your locality.

Bonding screw. The second important part of grounding your electrical system is to make sure the bonding screw is installed in the service entrance panel (see drawing **65-A**). The bonding screw bonds (electrically joins) the metal cabinet to the neutral bus bar, which is tied to earth through the grounding electrode conductor. This grounds all metal parts bonded to the service entrance panel.

TWO TYPES OF SERVICE ENTRANCE

A special exception in the National Electrical Code states the size of type THW and THWN conductors for both overhead and underground service. See **Tables IV** and **V** on page 42.

If you use type TW conductors for overhead service, you must use one size larger (one size smaller AWG number) than that indicated in **Tables IV** and **V**.

If your service installation uses conduit, you can find the minimum conduit size required for the size of the service entrance conductors in **Table VI** on page 42. For underground service the utility company will often specify the conduit size.

Overhead Service

Drawing **62-A** shows a typical overhead service. Your utility customer service representative will generally specify the location and the mast height, but here are some facts to keep in mind. The service entrance conductor drip loops must be at least 10 feet from the ground, and the service wires cannot come within 3 feet of doors, windows, or other openings. In addition, certain minimum clearances of the service drop above ground, sidewalks, driveways, or roadways can increase the minimum mast height.

The mast (or any other structure where the service drop fastens) must be strong enough to withstand all weather conditions. Again, your local inspector or utility company can advise you.

Conduit that penetrates the roof must be rigid steel or aluminum conduit. If the roofline is high enough that the service entrance head can be below

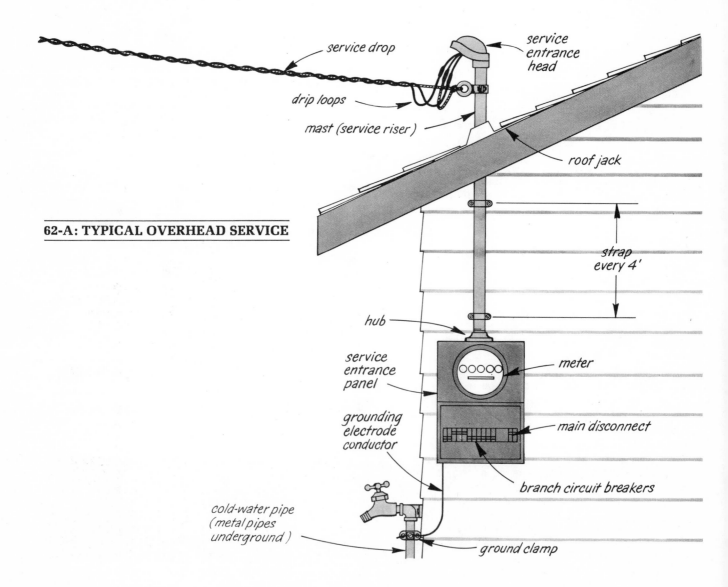

62-A: TYPICAL OVERHEAD SERVICE

the roofline, EMT may usually be used for the service riser (see drawing **63-A**). Check with your code enforcement agency about using nonmetallic conduit (PVC). Another variation allowed in some jurisdictions is the use of service entrance cable (SE), consisting of two insulated conductors wrapped with a stranded neutral (see page 37). Where necessary for connections, the neutral wire is formed by twisting the strands together. A special weather head is required when using SE cable.

Underground service

Drawing **63-B** shows a typical underground service. Depending upon local code requirements, you may be required to use rigid metal conduit for the underground riser, or you may be allowed to use an approved nonmetallic conduit (schedule 80 or even schedule 40).

In some cases the utility company may elect to bring in direct burial (DB) cables of its own instead of having conduit all the way to the pull box or transformer. Type UF is *not* permitted for service entrance use.

Note that the service entrance panel shown in drawing **63-B** includes an underground pull section. This space allows the utility company room to pull the service conductors through the underground riser.

63-A: ENTRANCE HEAD BELOW ROOFLINE

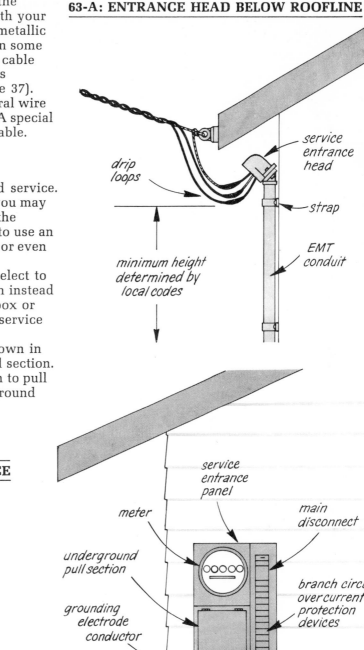

drip loops

service entrance head

strap

EMT conduit

minimum height determined by local codes

63-B: TYPICAL UNDERGROUND SERVICE

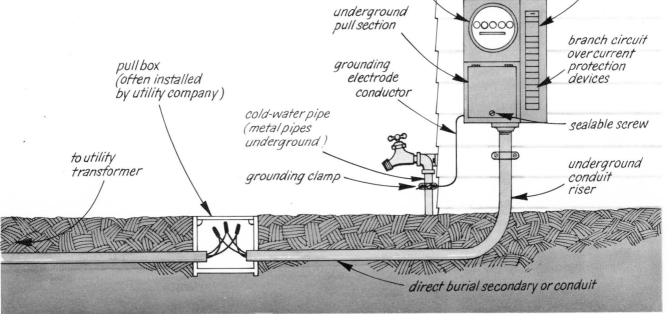

pull box (often installed by utility company)

to utility transformer

cold-water pipe (metal pipes underground)

grounding clamp

underground pull section

grounding electrode conductor

meter

service entrance panel

main disconnect

branch circuit overcurrent protection devices

sealable screw

underground conduit riser

direct burial secondary or conduit

Use of subpanels

Subpanels are sometimes advantageous, either for convenient access to overcurrent devices or to avoid many long circuit runs. You may wire all branch circuits from subpanels (in which case the service entrance panel contains only the main breaker, which is also the subfeed breaker) or the branch circuits may be divided among the main panel and subpanels.

Although it's a cabinet, a subpanel is used as a branch circuit. Like that of any other branch circuit, the subpanel's ampacity must not be less than that of the overcurrent device protecting it at the service entrance panel or in the subpanel itself. In residential construction, for example, a subpanel rated at 125 or 150 amps is often served by a subfeed cable rated at 100 amps and protected by a 90 or 100-amp breaker.

WIRING A SUBPANEL

The wires leading from the service entrance panel to a subpanel are called "subpanel feeders" or just "subfeeds." At the service entrance panel, the subfeeds connect to circuit breakers just like any other 120/240-volt branch circuits (see drawing 65-A). In most subpanels the subfeeds attach directly to the hot bus bars, requiring no subpanel main disconnect.

"Floating" neutral bus bar

Unlike the neutral bus bar in the service entrance panel, the subpanel neutral bus bar is *not* grounded with a grounding electrode conductor, and the bonding screw is not installed in a subpanel. The neutral bus bar is left "floating." Neutral bus bars, and therefore neutral wires, are grounded only at the service entrance panel.

Converting to a subpanel

A convenient way to increase your service is to install a new service entrance panel and equipment, and then to convert your present service entrance panel into a subpanel. To perform this conversion, you must remove the grounding electrode conductor and the bonding screw. In addition, if branch circuit grounding wires are present, you must remove them from the neutral bus bar and connect them to a grounding bus bar that is grounded to the neutral bus bar in the new service entrance panel. The original service entrance conductors are either removed completely or folded and taped out of the way. And finally, subfeeds from the new service entrance panel are connected as in any other subpanel.

To accomplish this conversion with the shortest interruption of your service, you'll have to coordinate carefully with the utility company and your electrical inspector.

> ⚠ Do not do anything with the original service entrance conductors until the utility company has severed them from incoming power.

Branch circuit connections at panels

The actual connection of branch circuits in the panels is usually best saved until all the circuits have been run. To keep things straight, it's a good idea to label the cable or wire ends in the panel as you put in the circuits.

For connecting branch circuits in a circuit breaker type panel, be sure to purchase the correct style of circuit breaker for your panel. You'll need to know the brand and approximate age of panel (if possible, get number of panel) in order to buy the correct breaker.

Each branch circuit requires three basic connections in a panel (whether a service entrance panel or a subpanel). The hot wire (or wires, in the case of 240-volt circuits), the neutral wire, and the grounding wire all have specific connection points. However, if you run your circuit wires inside metal conduit (EMT, IMC, or rigid conduit), you'll only be concerned with hot and neutral wires; a separate grounding wire is not usually necessary. The conduit itself provides the necessary path to ground through the bonding screw in the service entrance panel.

Because some of the connection points differ between service entrance panel and subpanel, we will discuss branch circuit connections at each type of panel separately.

CONNECTIONS IN THE SERVICE ENTRANCE PANEL

Branch circuits fall into three categories: 120-volt, 120/240-volt, and 240-volt circuits. Most circuits in your home are 120-volt circuits.

Some appliances, such as a range or a clothes dryer, require inputs of both 120 and 240 volts. Timing devices and motors in the appliances need 120 volts, and heating elements need the additional power of 240 volts.

Electric water heaters and central air conditioners are examples of appliances that require a straight 240-volt circuit.

120-volt circuits. Connecting 120-volt branch circuits at the service entrance panel is not difficult. Follow these three basic steps: 1) Connect the neutral wire (white or gray insulation) and grounding wire (bare or green insulation) directly to the bonded neutral bus bar. 2) Connect the hot wire (black or red insulation) to the overcurrent protection device. On a circuit breaker, the connection point is the screw terminal. Install the breaker (preferably before the circuit wire is attached) on a hot bus bar by pushing until the clip on the back of the breaker is firmly fastened to the bus. For fused panels, the connection point is the screw terminal next to the fuseholder. 3) Balance your total load. Be sure you have approximately the same amount of load on each hot bus bar.

120/240-volt circuits. Connecting 120/240-volt branch circuits is exactly the same as hooking up 120-volt circuits with the exception that two hot wires are used along with a neutral. These wires connect, through the overcurrent protection device, to *each* of the two hot bus bars (see drawing **65-A**). With circuit breakers, always use a common-trip, two-pole breaker with one handle when wiring a 120/240-volt circuit. This way, if one hot wire is accidentally grounded, the circuit will be disconnected from both hot bus bars.

240-volt circuits. A strictly 240-volt circuit consists of two hot wires (one from *each* hot bus bar) and a grounding wire. The complete loop of the circuit is formed by the two hot wires. At any given instant, these wires have opposite voltage polarities (as though one had positive pressure and the other negative pressure or suction); therefore they form a current loop.

In the service entrance panel, connect one hot wire to each of the hot bus bars and connect the

65-A: SERVICE ENTRANCE PANEL (OVERHEAD OR UNDERGROUND)

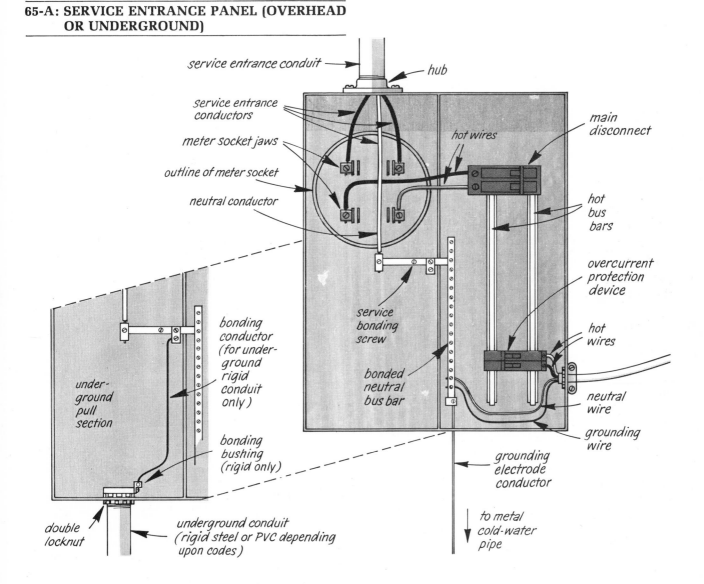

service entrance conduit · hub · service entrance conductors · meter socket jaws · outline of meter socket · neutral conductor · hot wires · main disconnect · hot bus bars · overcurrent protection device · hot wires · service bonding screw · bonded neutral bus bar · neutral wire · grounding wire · underground pull section · bonding conductor (for underground rigid conduit only) · bonding bushing (rigid only) · double locknut · underground conduit (rigid steel or PVC depending upon codes) · grounding electrode conductor · to metal cold-water pipe grounding wire to the neutral bus bar.

Multiwire circuits. "Split circuit" or multiwire circuit wiring is a special use of a 120/240-volt circuit. The 120/240-volt circuit runs to a junction box, where it is split into two 120-volt circuits. This is shown schematically in drawing **66-A**.

66-A: MULTIWIRE CIRCUIT

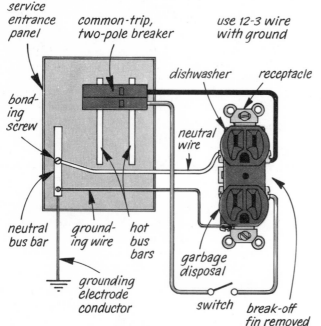

service entrance panel

common-trip, two-pole breaker

use 12-3 wire with ground

dishwasher   receptacle

bonding screw

neutral wire

neutral bus bar

grounding wire

hot bus bars

garbage disposal

grounding electrode conductor

switch

break-off fin removed

This method of wiring is useful, for example, when you want to run a single three-wire cable with ground (three current-carrying conductors and one grounding wire) instead of two two-wire cables with ground (four current-carrying conductors and two grounding wires). One use of this technique is for wiring circuits for a dishwasher and a garbage disposal.

At the service entrance panel, you can connect the hot wires to a common-trip two-pole (240-volt) breaker. As with all other branch circuits at the service entrance panel, the neutral and grounding wires are connected to the bonded neutral bus bar.

BRANCH CIRCUIT CONNECTIONS TO SUBPANEL

Drawings **66-B** and **66-C** show two typical subpanels. The drawings show that the hot wires connect to the overcurrent protection devices just as they do in the service entrance panel.

Grounding wires in a subpanel are either connected to their own bonded bus bar (see drawing **66-B**) or cinched together with a compression lug and bonded (screwed) to the panel (see drawing **66-C**).

SIZING YOUR BRANCH CIRCUITS

Choosing the proper conductor size for a branch circuit is not difficult. First determine the amp rating necessary for the circuit, and then look up the appropriate wire size in **Table IV** or **Table V** (page 42), depending on whether you're using copper, aluminum, or copper-clad aluminum conductors.

Another factor enters the picture if you intend to run more than three current-carrying wires (this excludes grounding wires) in a confined space such as conduit. In such a case the wires must be derated — their ampacity rating must be lowered — to prevent a detrimental buildup of heat. This means in effect that a larger wire size will be required for circuits of a given capacity. Your local inspector can give you details.

The amperage rating of your overcurrent protection device must be less than or equal to the amp rating of your circuit wires (the derated amperage if any derating was necessary). This is important because you want the device to blow or trip if more than the rated amount of current flows in the wires.

66-B: GROUNDING WIRES IN SUBPANEL

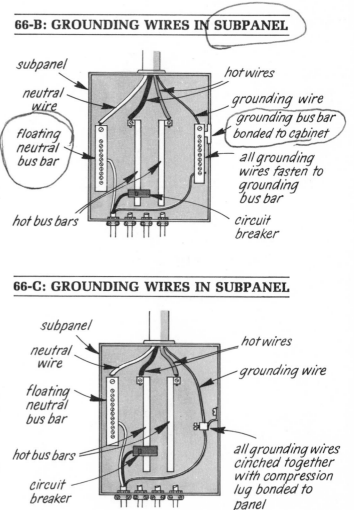

subpanel

neutral wire

floating neutral bus bar

hot bus bars

hot wires

grounding wire

grounding bus bar bonded to cabinet

all grounding wires fasten to grounding bus bar

circuit breaker

66-C: GROUNDING WIRES IN SUBPANEL

subpanel

neutral wire

floating neutral bus bar

hot bus bars

circuit breaker

hot wires

grounding wire

all grounding wires cinched together with compression lug bonded to panel

Standard 120-volt branch circuits. Most circuits in your home are 15 or 20-amp general purpose circuits and 20-amp small appliance circuits. For the 15-amp circuits use #14 copper wire; for the 20-amp circuits use #12 copper wire. (For other wire types, see **Table V**, page 42.)

Major appliance circuits. When planning the individual circuit for a major appliance (such as a range, dryer, air conditioner, or water heater), check the appliance installation information for the necessary circuit size.

Here are typical circuit sizes for some appliances.
- Electric dryer: 30 amp, 120/240 volt, #10 copper wire.
- Range: 50 amp, 120/240 volt, #6 copper wire.
- Dishwasher: 20 amp, 120 volt, #12 copper wire.
- Garbage disposal: 20 amp, 120 volt, #12 copper wire.

WIRING A 120/240-VOLT APPLIANCE RECEPTACLE

There are two ways to wire a 120/240-volt appliance receptacle, depending on whether the circuit runs from the main service entrance panel or a subpanel. Drawing **67-A** shows these two situations with a clothes dryer receptacle.

67-A: 120/240-VOLT RECEPTACLE

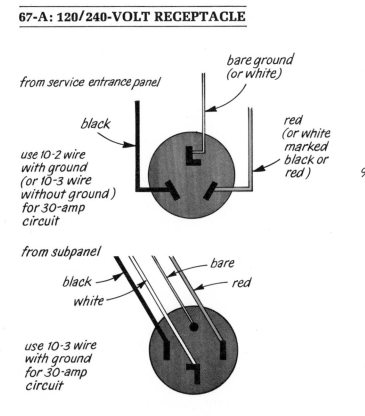

from service entrance panel

bare ground (or white)

black

red (or white marked black or red)

use 10-2 wire with ground (or 10-3 wire without ground) for 30-amp circuit

from subpanel

bare

black

red

white

use 10-3 wire with ground for 30-amp circuit

REQUIRED DISCONNECTS

Some stationary, motor-operated appliances (such as a garbage disposal, the motor of a gas heater, or a central air conditioner) require separate disconnects. Check with your local inspector before installing.

CONNECTING CONDUIT TO PANELS

Conduit is fastened to the service entrance panel or a subpanel the same way it's connected to any box. Choose a connector that's appropriate for the location, type, and size of conduit you'll be using. Punch out a knockout appropriate for the conduit and fasten the conduit securely to the panel with the connector. Drawing **67-B** shows several typical connectors and arrangements for conduit. (For more information on working with conduit, see pages 74-78.)

Once your conduit system is installed and you are pulling your circuit conductors through, be sure to leave plenty of wire extending into the panel: 18 inches is not too much for wiring a large panel.

67-B: CONDUIT CONNECTORS

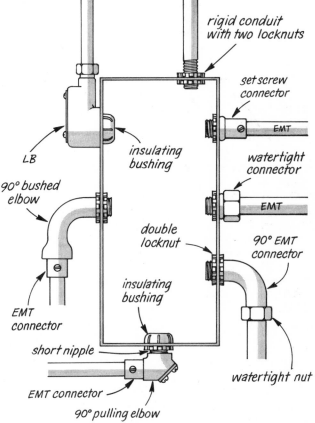

rigid conduit with two locknuts

set screw connector

EMT

watertight connector

EMT

90° EMT connector

insulating bushing

double locknut

insulating bushing

watertight nut

LB

90° bushed elbow

EMT connector

short nipple

EMT connector

90° pulling elbow

CONNECTING NONMETALLIC SHEATHED CABLE TO PANELS

Where nonmetallic sheathed cable enters a service entrance panel or a subpanel, you have some options. After removing the correct size knockouts for your cable, you may either use a cable connector or a snap-in plastic clamp to hold the cable (see drawing **68-A**). The plastic clamp snaps into the knockout hole to prevent chafing; the cable must then be stapled within 12 inches from the box.

Be sure to leave plenty of cable extending into the panel to be stripped back and connected to the overcurrent protection devices: 18 inches isn't too much for wiring a large panel.

68-A: NM CABLE CONNECTORS

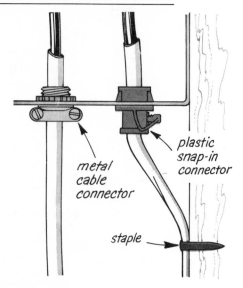

metal cable connector

plastic snap-in connector

staple

Wiring with nonmetallic sheathed cable

Nonmetallic sheathed cable, conduit, armored cable, and knob-and-tube (not approved for new installations, but may be used as an extension of existing wiring) represent four basic methods for wiring branch circuits. Today nonmetallic sheathed cable is used in most resdental wiring. This cable is a good example of a successful marriage of modern materials and technology. The nonmetallic sheathed cable comes in types NM (nonmetallic) and UF (underground feeder).

WHERE IT CAN BE USED

Both type NM and type UF cables can be used for exposed (new) or concealed (old) work, though UF is usually used as a direct burial cable in the earth. Type NM is allowed only in dry locations; UF may be used in either dry or wet locations. Neither cable may be used as service entrance cable nor embedded in cement, concrete, aggregate, or plaster. Both cables may run in the air spaces of masonry block or tile walls — NM if not damp, UF if damp.

HOW AND WHERE TO RUN NONMETALLIC SHEATHED CABLE

Try to not kink or twist NM or UF cable during installation. Wherever possible, run cable along the surface of structural members (studs, joists, rafters, etc.). Drill holes and run cable through the boards where you must run at an angle to them (see drawing **69-A**).

> ⚠ Don't route cables where you're likely to accidentally nail through them later.

Estimating length of cable. Make a rough sketch of your cable route, including critical dimensions such as height of boxes from floor and distance between boxes. Add together all dimensions. Then, for box-to-box runs, add 4 feet (2 feet for each box) for mistakes, box connections, and unforeseen obstacles. When cable goes from a panel (either service entrance or subpanel), add 6 feet (4 for the panel and 2 for the box).

Drilling holes. Use the smallest size drill possible, such as ¾ inch, when boring through joists or studs. To avoid weakening the wood members excessively, drill exactly in the center of the board. If your hole is less than 1½ inches from the edge of the board, tack a metal plate or a piece of 16-gauge metal over the edge (see drawing **69-A**).

Supporting cable. In exposed (new) wiring, cable must be stapled or supported every 4½ feet and within 12 inches of each metal box and 8 inches of each nonmetallic box (see drawing **69-A**). Cable staples or supports are not required when cable is fished behind walls, floors, or ceilings in concealed (old) work.

Unfinished basement wiring

When run under the floor at an angle to floor joists, NM with two conductors in sizes smaller than #6 or with three conductors in sizes smaller than #8 must be either bored through the joists, secured to running boards, or supported on the surface of structural members (see drawing **70-A**). Larger NM cable may be stapled directly to the bottom edges of the joists.

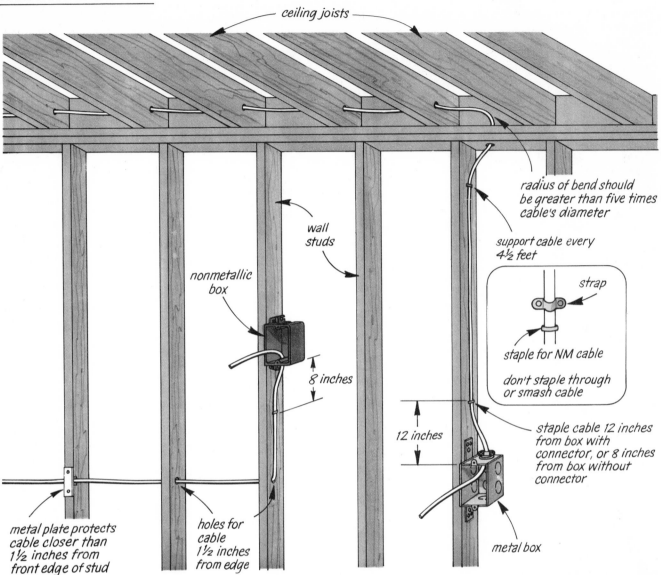

ceiling joists

radius of bend should
be greater than five times
cable's diameter

wall
studs

support cable every
4½ feet

nonmetallic
box

strap

staple for NM cable

don't staple through
or smash cable

8 inches

staple cable 12 inches
from box with
connector, or 8 inches
from box without
connector

12 inches

metal plate protects
cable closer than
1½ inches from
front edge of stud

holes for
cable
1½ inches
from edge

metal box

Attic wiring

Accessibility dictates how cable runs in an attic. If a permanent staircase or ladder leads to the attic, cable running at an angle to structural members must be protected by guard strips. In an attic reached through a crawl hole with no permanent stairs or ladder, the cable must be protected by guard strips only within 6 feet of the hole (see drawing **70-B**). Beyond the 6 feet, cable may lay on top of the ceiling joists.

NM CABLE ENTERING BOXES

In wiring with nonmetallic sheathed cable, you may use either metal or nonmetallic boxes. It would be difficult to recommend one type over the other because both have advantages. Several aspects of wiring, such as grounding and connecting cable to boxes, do vary depending on the kind of box you use. For this reason we'll consider metal and nonmetallic boxes separately.

Metal boxes

Cable must be secured to a metal box. This may be done by using either a box with built-in cable clamps (see drawing **71-A**) or a separate cable connector (see drawings **71-B** and **71-C**). Either way, remember to leave 6 to 8 inches of cable extending into the box for connections.

70-A: NM CABLE UNDER FLOOR

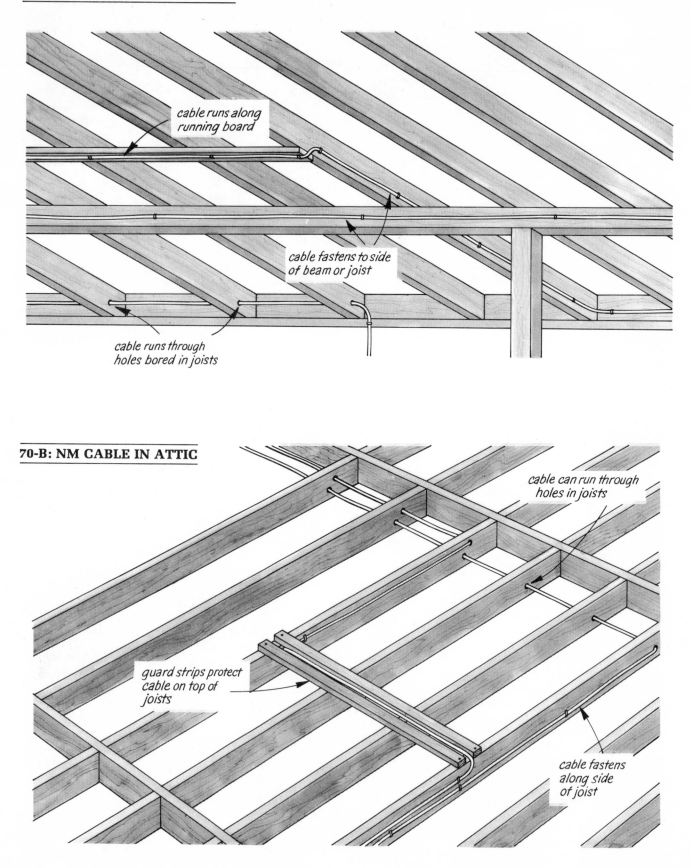

cable runs along running board

cable fastens to side of beam or joist

cable runs through holes bored in joists

70-B: NM CABLE IN ATTIC

cable can run through holes in joists

guard strips protect cable on top of joists

cable fastens along side of joist

71-A: BUILT-IN CABLE CLAMP

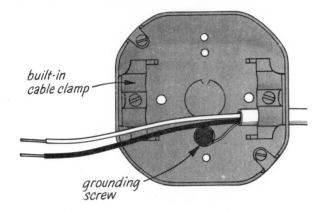

built-in cable clamp

grounding screw

As with most other terminal connectors, a grounding screw or a grounding clip is equipped to receive only one wire. When you install a switch, self-grounding receptacle, or light fixture in the *last* box of a circuit run, attach the grounding wire of NM cable directly to the box as shown in drawings **71-A** and **71-D**.

When you install a box that's not at the end of a circuit or a non-self-grounding receptacle, you will have to make up one or more grounding jumpers. For the grounding jumper, use wire the same size as the circuit wires. Twist all grounding wires and jumpers together and crimp with a compression ring or secure with a wirenut (see drawing **71-E** and pages 44 and 45).

71-B: METAL CABLE CONNECTOR

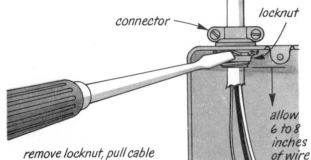

connector

locknut

allow 6 to 8 inches of wire

remove locknut, pull cable through connector, tighten screws, push connector through knockout, slide on locknut and tighten with a screwdriver

71-D: USING A GROUNDING CLIP

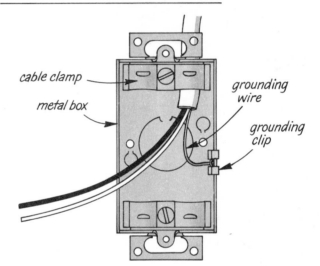

cable clamp

metal box

grounding wire

grounding clip

71-C: PLASTIC CABLE CONNECTOR

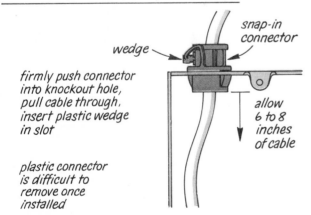

wedge

snap-in connector

firmly push connector into knockout hole, pull cable through, insert plastic wedge in slot

allow 6 to 8 inches of cable

plastic connector is difficult to remove once installed

Grounding metal boxes. Metal boxes must be grounded. A separate grounding wire in NM cable provides the path for ground; a screw (drawing **71-A**) or grounding clip (drawing **71-D**) bonds the box to the wire.

71-E: GROUNDING JUMPER

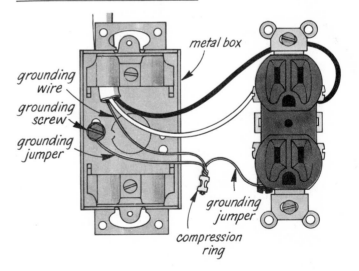

metal box

grounding wire

grounding screw

grounding jumper

grounding jumper

compression ring

Nonmetallic boxes

Cable knockouts in nonmetallic boxes are held in place with thin webs of plastic. Break out a knockout wherever NM cable will enter a box. NM cable need not be clamped to a nonmetallic box if it is stapled within 8 inches of the box.

In old work, where you're fishing cable behind walls, ceilings, and floors, you may not be able to support a cable within 8 inches of a box. In this case you must clamp the cable to the box. "Old work" nonmetallic boxes are available with built-in cable clamps (see page 80).

Grounding and nonmetallic boxes. Since nonmetallic boxes don't conduct electricity, they need not be grounded. Then what should be done with the grounding wire in NM cable when you use nonmetallic boxes? The grounding wire is treated a little differently according to whether the box holds a receptacle, switch, or light fixture.

A receptacle in a nonmetallic box is no problem. If the box is at the end of the circuit, the grounding wire attaches to the grounding screw of the receptacle (see drawing **72-A**). If the box is in the middle of a circuit run, the cable grounding wires are joined with a grounding jumper from the receptacle (see drawing **72-B**).

Most switches don't have grounding terminals. If you have a switch in a nonmetallic box in the middle of a run, simply join together the grounding wires of the cables (see drawing **72-C**). If your switch and box are at the end of a circuit, place the grounding wire between the switch bracket and the box. Then be sure to fasten the mounting screw tightly (see drawing **72-D**).

To install a light fixture, use a round fixture box with a metal grounding bar. If the fixture is at the end of the circuit, attach the cable grounding wire to the bar (see drawing **48-C**). If the fixture is in the middle of a circuit, make up a grounding jumper to join the grounding bar to the cable grounding wires. The light fixture will automatically be grounded when attached to the grounded box. (Some chain-hung fixtures have a separate grounding wire. Join it to the circuit grounding wire and a jumper as shown in drawing **48-B**.)

72-A: END OF CIRCUIT

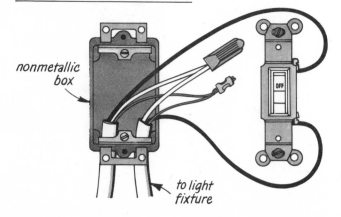

72-B: MIDDLE OF CIRCUIT

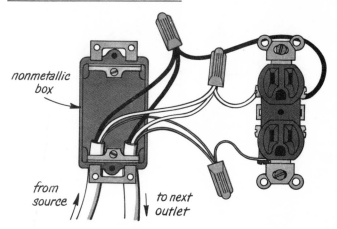

72-C: MIDDLE OF CIRCUIT

72-D: END OF CIRCUIT

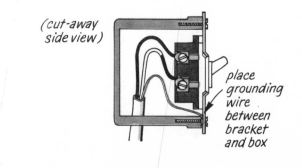

Wiring with armored cable

Type AC metal-clad cable is very expensive; as a result, it is seldom used today for residential wiring. If you do use it, take care to learn to use it correctly. See your local inspector.

WHERE IT CAN BE USED

Type AC cable may be used for branch circuits and subfeeds of both exposed (new) and concealed (old) wiring in dry locations only. AC cable may be run in the air spaces of masonry block or tile walls if they are dry and above ground level. You cannot bury type AC cable in the earth or in concrete, but you can embed it in a plaster finish over a brick or other masonry wall.

HOW AND WHERE TO RUN AC CABLE

Wherever possible, route your circuit so that the cable follows the surface of structural members. Where you must run the cable at an angle to the members, drill holes in the center of boards. A metal plate must be tacked to any board where a hole is drilled less than 1½ inches from the edge (see drawing **69-A**).

Though the metal armor of AC cable is flexible, sharp bends can be damaging. The bend radius should be at least five times the diameter of the cable.

How to cut AC cable. Cut AC cable cautiously with a fine-toothed hacksaw. Rotating the cable as you go, cut only the outer skin of metal. After removing the excess metal armor, insert a special fiber or plastic protective bushing in the cut end to prevent chafing of wire insulation (see drawing **73-A**).

73-A: AC CABLE

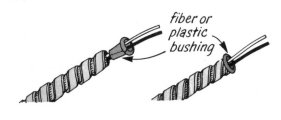

fiber or plastic bushing

Supporting cable

Use only staples or straps designed for use with AC cable. Type AC cable must be supported every 4½ feet and within 12 inches of each box (see drawing **69-A**).

Unfinished basement/attic wiring. When routing type AC cable at an angle to floor joists, attach the cable directly to each joist (see drawing **73-B**). When routing in attics, treat AC cable like NM (see drawing **70-B**).

73-B: AC CABLE UNDER FLOOR

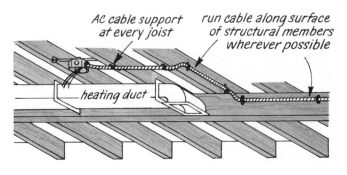

AC cable support at every joist — run cable along surface of structural members wherever possible — heating duct

AC CABLE ENTERING BOXES

You can only use metal boxes when wiring with type AC cable. At each termination, the cable must be secured to the box. Either use a separate connector like the one shown in drawing **73-C**, or use a box with built-in clamps especially designed for AC cable, also shown in drawing **73-C**. Both the connectors and cable clamps have slots so that the protective bushing is visible once the installation is completed.

73-C

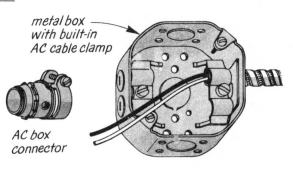

metal box with built-in AC cable clamp

AC box connector

Grounding and AC cable. A special bonding strip is attached to the inside of the metal armor of all AC cable. This bonding strip provides the grounding path for AC cable wiring. Because of this, a separate grounding wire is not necessary when the cable is connected to a metal box. A jumper is still required, however, between the grounding terminal of a receptacle and the metal box.

Wiring with conduit

An important point to remember if you choose to wire with conduit is that you must install the entire conduit system before pulling the conductors through. Keep this in mind as you design your conduit system. Allow enough suitably placed fittings to ensure that the conductor pulls will be as direct as possible.

If your conduit run contains more than 360° in *total* bends (such as four 90° bends or three 90° plus two 45° bends), you should use a pull box. This is an intermediate box used only for pulling and connecting wires. It is then capped with a blank cover. Drawing **74-A** shows a square junction box with a single-device plaster ring used as a pull box. After the wallboard is installed and wires are pulled, a blank faceplate should be added.

For exposed conduit, an alternative is to use corner pulling elbows that break apart for pulling, and then are sealed with screws (see drawing **74-B**). The 360° rule does not apply if these fittings are used, but still applies to all other bends.

NONFLEXIBLE METAL CONDUIT

Thinwall conduit (EMT), intermediate metal conduit (IMC), and rigid metal conduit (rigid) are three basic types of nonflexible metal conduit. They represent a three-stage graduation in strength. Many installation techniques are the same for all three, though certain applications vary as the conduit strength increases.

Sizing metal conduit. To find out what size conduit you need for the number of conductors you'll be using, see **Table VI**, page 42.

Grounding a metal conduit system

One important feature of all nonflexible metal conduit is that it provides the grounding path back to the neutral bus bar in the service entrance panel. It is not necessary to run a separate grounding wire with the circuit conductors. To maintain the grounding continuity, all couplings, connectors, fittings, and boxes must be metal and all connections must be made tight.

74-A: CONCEALED CONDUIT

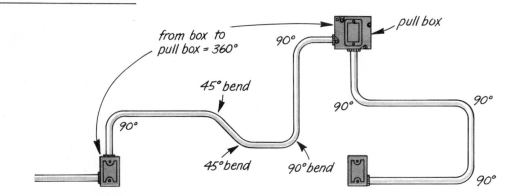

74-B: SURFACE CONDUIT

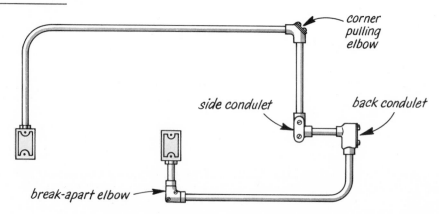

Where to use nonflexible metal conduit

Nonflexible metal conduit may be used inside and out, in wet and dry locations, and for exposed and concealed wiring. When conduit is used in wet locations, all couplings, connectors, and fittings must be watertight. EMT should not be used underground. However, some types of IMC and rigid may be buried directly in the earth or embedded in concrete.

How to use nonflexible metal conduit

Metal conduit is sold in 10-foot sections. To get the exact length and configuration, your conduit will require cutting, bending, and coupling.

Cutting and reaming. A hacksaw is all you need to cut conduit. But sawing isn't all there is to a conduit cut. You must ream each cut end to remove all burrs and sharp edges that could damage conductor insulation. Give a few quick turns around the inside of the cut with a round metal file.

Using a regular pipe threader, you may rethread IMC and rigid conduit after cutting. Most homeowners, however, find it more practical to use special threadless connectors and couplings instead of rethreading.

75-A: HOW TO BEND CONDUIT

90° bend

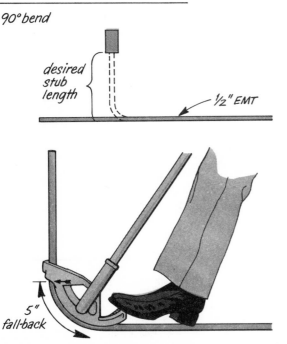

Step 1. *Check bender for 90° stub fall-back measurement (usually 5 inches for ½-inch EMT). Subtract fall-back from finished stub length and mark conduit.*

Step 2. *Slide bender over conduit and align arrow on bender with mark on conduit. Bend up stub by stepping on bender pulling handle toward you.*

Bending. All conduit bends must be made in such a way that the internal size of the conduit is not reduced. This means that you should use a conduit bender. Drawing **75-A** shows how to measure for bends and how to use a bender.

back-to-back bends

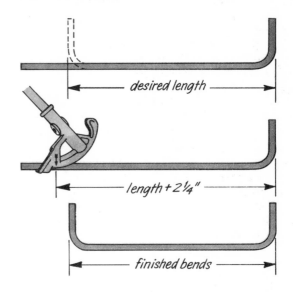

Step 1. *Make first stub bend and measure length of conduit you'll need.*

Step 2. *Check bender for back-to-back overshoot (usually 2¼ inches for ½-inch EMT). Add overshoot to desired length.*

Step 3. *Bend second stub back toward first stub.*

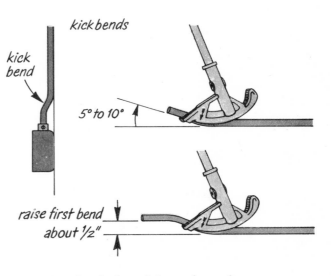

Step 1. *Bend end of conduit up 5° to 10°.*

Step 2. *Turn conduit 180°, move forward in bender, and bend again. Second bend should raise first bend about ½ inch, depending on box.*

Design your runs so that no piece of conduit between two boxes or fittings bends more than the equivalent of four quarter turns (see drawing **76-A**).

76-A: FOUR QUARTER TURNS

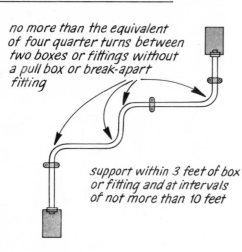

no more than the equivalent of four quarter turns between two boxes or fittings without a pull box or break-apart fitting

support within 3 feet of box or fitting and at intervals of not more than 10 feet

Coupling. Conduit is sold in 10-foot lengths, so you'll have to join sections for long conduit runs. IMC and rigid conduit are threaded on the ends and are sold with one coupling for each length. (IMC and rigid couplings are interchangeable.) Lengths of these types of conduit screw together just like water

pipe. EMT conduit, on the other hand, has walls too thin to thread. Threadless couplings (which fit only EMT) are used to join sections of EMT conduit. Various types of EMT couplings are available for indoor and outdoor use (see drawing **76-B**).

Joining conduit to boxes. For indoor use, conduit joins a metal box through a knockout. You can't join conduit to a round box; use octagonal and square boxes instead. Be sure the boxes you get have knockouts large enough to accommodate the size of conduit you're using.

With IMC and rigid conduit, the box is secured on the threaded end of the conduit between two

76-C: IMC & RIGID FITTINGS

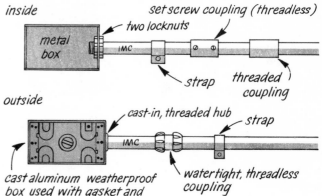

inside

metal box

IMC

two locknuts

set screw coupling (threadless)

strap

threaded coupling

outside

cast aluminum weatherproof box used with gasket and cover

cast-in, threaded hub

strap

IMC

watertight, threadless coupling

76-B: EMT FITTINGS

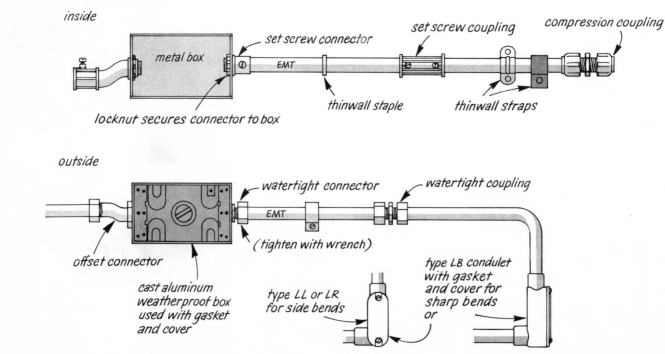

inside

metal box

EMT

set screw connector

set screw coupling

compression coupling

thinwall staple

thinwall straps

locknut secures connector to box

outside

watertight connector

watertight coupling

EMT

(tighten with wrench)

offset connector

cast aluminum weatherproof box used with gasket and cover

type LL or LR for side bends

type LB condulet with gasket and cover for sharp bends

or

locknuts (see drawing **76-C**). With EMT, an EMT set-screw connector can be used (see drawing **76-B**).

For outdoor work, use a cast aluminum weatherproof box. The correct box for IMC and rigid conduit has a cast-in threaded hub so that the conduit can be screwed in directly (see drawing **76-C**). For EMT, use a watertight connector which fastens to a threaded boss on the box (see drawing **76-B**).

FLEXIBLE METAL CONDUIT

Almost always used in dry locations, flexible metal conduit ("flex") cannot be used underground, or embedded in concrete or aggregate. If, however, wet-location-approved conductors are used, you may use flex in wet places.

Grounding flexible conduit

Because of restrictions, most flexible conduit systems are grounded by running a separate grounding wire along with the circuit conductors.

Bending and supporting flexible conduit

Despite its flexibility, runs of flexible conduit between boxes and fittings must not bend more than the equivalent of four quarter turns (see drawing **76-A**). Flex must be supported with a conduit strap within 12 inches of every box or fitting and at intervals not longer than 4½ feet.

Drawing **77-A** shows several examples of flexible conduit connectors.

RIGID NONMETALLIC CONDUIT

There are many types of nonmetallic conduit, but schedule 40 PVC is the one most homeowners wish to use. It is a rigid, heavy-walled, flame-retardant, heat and sunlight-resistant conduit. It may be used in wet or dry locations, in walls, ceilings, and floors, and above or below ground. Do not try to substitute PVC irrigation pipe for schedule 40 PVC (look for the insignia of an electrical materials testing laboratory).

You can use nonmetallic conduit with either metal or nonmetallic boxes, but the nonmetallic boxes are not the same as used for NM cable.

Wiring with PVC

Nonmetallic conduit does not constitute a grounded system, so you must run a separate grounding wire with the circuit conductors.

Trimming. After cutting PVC, trim the ends—inside and out — with a pocket knife to remove any rough edges that might damage conductor insulation.

Bending. Bends in PVC are made by heating the conduit in a special infrared heater until it is soft. Don't try to heat PVC with a torch; you'll just char the conduit.

Design your runs so that no piece of conduit between two boxes or fittings bends more than the equivalent of four quarter turns (see drawing **76-A**).

Joining PVC. PVC comes in 10-foot lengths, each with one coupling. Glue the conduit together with gray conduit cement. Don't use water pipe cement.

Male and female adapters are available for transitions to other types of conduit and for box connections. Drawing **77-B** shows the use of PVC adapters.

77-A: FLEX CONNECTIONS

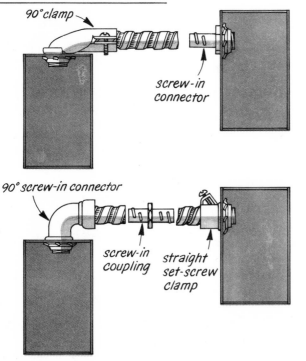

90° clamp

screw-in connector

90° screw-in connector

screw-in coupling

straight set-screw clamp

77-B: PVC ADAPTERS

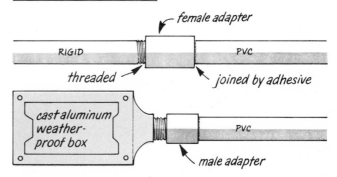

female adapter

RIGID

PVC

threaded

joined by adhesive

cast aluminum weather-proof box

PVC

male adapter

Supporting PVC. Supports for nonmetallic conduit should be placed within 4 feet of each box or fitting. In most instances, additional supports should be placed at least every 4 feet.

Burying PVC. PVC that's approved for direct burial in the earth without a cement encasement must be buried at least 18 inches deep. If you put a 2-inch concrete pad over the conduit, you may bury it as little as 12 inches deep.

PULLING WIRES IN CONDUIT

Once you have installed your conduit and boxes, you need to pull the wires through.

If you are pulling a few #10, #12, or #14 wires, pull directly with a fish tape. Push (unreel) the tape through the conduit until it is exposed at the other end. Strip off 2 or 3 inches of insulation from the end of each wire and bend them tightly over the fish tape loop. Then pull the wires through the conduit by rewinding the fish tape (see drawing **78-A**).

78-A

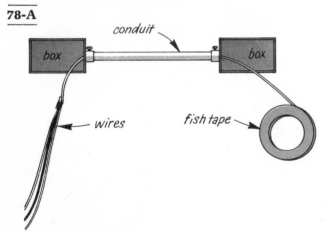

Step 1. *Push fish tape through conduit.*

Step 2. *Strip off 2 to 3 inches of insulation then bend wires tightly over fish tape's hook. Tape hook and apply pulling lubricant for pulling through curves.*

Step 3. *Rewind fish tape, pulling wires through conduit.*

To avoid kinking or scraping insulation, have someone feed the wires in as you pull. In many instances it also helps to have your helper apply pulling lubricant while feeding the wires into the conduit. Similar in consistency to waterless soap, pulling lubricant is a special compound that makes the conductors slide more freely yet is compatible with electrical insulation.

If you must do the pulling alone, precut all wires to the conduit length plus 2 feet. Lay the wires in a straight line from the end of the conduit so the pull will be as direct and easy as possible.

Knob-and-tube wiring

For this wiring method, individual conductors are supported and protected with nonmetallic knobs, tubes, and flexible tubing. No longer used in new construction, knob-and-tube wiring may be used only as an extension to existing knob-and-tube installations.

One recommended way to extend a knob-and-tube circuit is to use NM cable. Drawing **79-A** shows how to tap into a knob-and-tube circuit for an extension using NM cable.

How to mount boxes

Where you're going to locate a box is a question that deserves some thought. Every box — whether it's a switch, outlet, junction, or pull box — must be accessible.

Except in kitchens, don't put boxes for receptacles too low or too high. Keep them between 12 and 18 inches from the floor. Avoid putting a switch box behind a door or on the hinge side. If you are wiring new construction, put either the top or the bottom of a switch box 48 inches from the floor. That way only one piece of wallboard will have to be cut instead of two.

Knockouts. Cable enters metal boxes through prestamped knockout openings. To remove a knockout, hold a screwdriver (or similar tool) on it and then strike the screwdriver with a hammer. Using a pair of pliers, twist off the punched-out knockout.

After your boxes are in place, you may discover that you removed too many knockouts or the wrong ones. Drawing **78-B** shows two types of seals that fit the round knockouts on metal boxes.

78-B: KNOCKOUT SEALS

snap-in cover

bar-type cover

79-A: EXTENDING KNOB-AND-TUBE

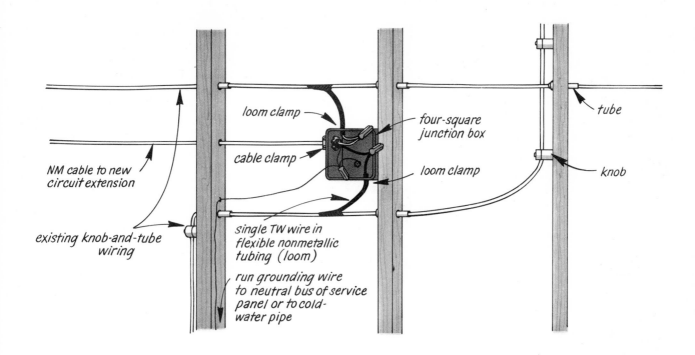

loom clamp

four-square
junction box

tube

cable clamp

loom clamp

knob

NM cable to new
circuit extension

existing knob-and-tube
wiring

single TW wire in
flexible nonmetallic
tubing (loom)

run grounding wire
to neutral bus of service
panel or to cold-
water pipe

Mounting arrangements. Whether a box is metal or nonmetallic, there are four main mounting methods: internal screw, external nail or bracket, hanger bar, and grippers.

Drawing **79-B** is a sampling of nonmetallic boxes showing these various mounting arrangements.

There are about as many kinds of external brackets as there are situations in which boxes are mounted. Before you buy a box, consider where you're going to mount it. You can save a lot of time by getting a box with brackets to suit your particular mounting needs.

If you want to hang an outlet box between two joists but don't have a box with a hanger bar, you can nail a 2 by 4 between the joists and attach the box directly to the wood.

Aligning boxes. Mount your boxes so they will be flush with the finished wall surface. How do you do this when the wall surface isn't in place when you install the boxes? One answer is to make a simple gauge out of your wall covering material. If, for instance, your walls and ceilings will be covered with wallboard, put a nail through a scrap of wallboard. Tack this scrap to the stud or joist next to your box and use it as a guide.

Note. Nonmetallic boxes are breakable. When mounting one with nails, be careful to hammer only the nails, not the box.

79-B: NONMETALLIC BOXES

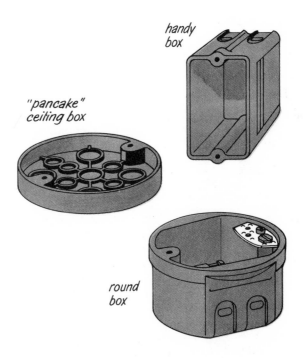

handy
box

"pancake"
ceiling box

round
box

(Continued on next page)

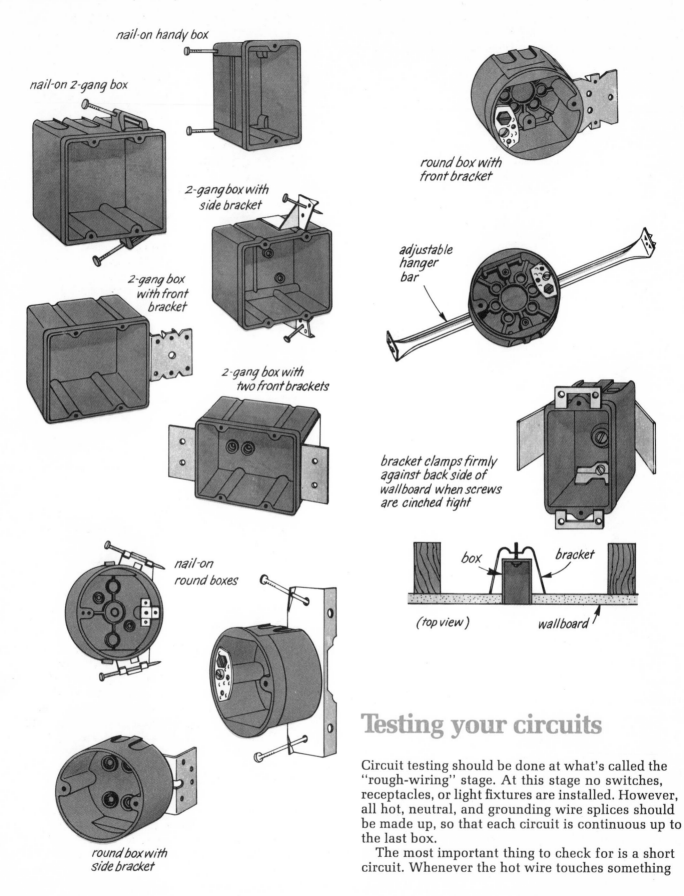

nail-on handy box

nail-on 2-gang box

2-gang box with
side bracket

2-gang box
with front
bracket

2-gang box with
two front brackets

nail-on
round boxes

round box with
side bracket

round box with
front bracket

adjustable
hanger
bar

bracket clamps firmly
against back side of
wallboard when screws
are cinched tight

box   bracket

(top view)   wallboard

Testing your circuits

Circuit testing should be done at what's called the
"rough-wiring" stage. At this stage no switches,
receptacles, or light fixtures are installed. However,
all hot, neutral, and grounding wire splices should
be made up, so that each circuit is continuous up to
the last box.

The most important thing to check for is a short
circuit. Whenever the hot wire touches something

grounded (such as the neutral wire, a grounding wire, or a grounded metal box), you'll have a short circuit. This could happen because insulation is damaged, a staple has penetrated cable, wiring is incorrect, or for some other reason. You should first check for shorts while the wiring is still exposed, because problems are easier to spot and correct. Test again after installing walls, ceilings, and floors but before final touches, because occasionally a nail will penetrate the wiring and cause a short.

If you want to take the time, you can also test for an open circuit. However, an open circuit is usually the result of an incorrect, loose, or missing connection at a box; this can be spotted and corrected quickly once the device is installed and the service is turned on.

WHAT TO TEST WITH

For circuit testing you can purchase a continuity tester (see page 36) or you can easily make your own. Here's how.

Tape a doorbell to a 6-volt dry cell battery. With a short piece of wire, connect one battery terminal to one doorbell terminal. Next, connect another wire about 2 feet long to the other terminal on the bell (see drawing **81-A**). To make sure you've done this correctly, test the tester by touching the free wire end to the free battery terminal momentarily. The doorbell should ring.

The idea of these testers is quite simple. A battery provides the power source and a light (if a continuity tester) or doorbell (if a homemade tester) provides a signal when the circuit is complete.

HOW TO TEST A CIRCUIT

It's a good idea to check your wiring visually first. Starting at the distribution center, walk through each circuit looking for problems. Think about what each wire is supposed to do; then verify that that's how you wired the circuit.

Testing for short circuits. So that your test will run the entire length of a circuit, temporarily join the hot wires each place you'll have a switch. This simulates the ON position and thereby extends your circuit test to include the wiring from the switch to whatever it will control.

At the service entrance panel, test your circuits as follows: Hook one lead of your tester (the alligator clip of a continuity tester or the free wire of a homemade tester) to the neutral bus bar; then touch the hot wire of each circuit to the tester (the tester probe or the free battery terminal) one at a time. The circuits should, of course, check out as

open (no bell or light). If the light comes on or the bell rings, you have a short in that circuit.

When testing circuits at a subpanel, you must run the same test described above twice. First, hook your clip to the neutral bus bar to check the hot wire/neutral wire circuits. Second, hook your clip to the grounding wire terminal to check your fault-current circuits.

Testing for open circuits. As we have said earlier, testing for open circuits can generally wait until power is turned on. However, if your wiring involves conduit that will be covered, you may want to check the continuity of your ground fault circuit (the bonding of the conduit sections). One method for doing this at each box is to touch the neutral wire and the box to the two ends of your tester. The light should come on or the bell should ring, indicating continuity. If it does not, check for a loose connection along the neutral or conduit path.

81-A: BELL TESTER

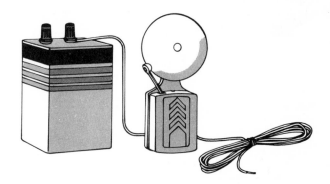

Tracking down a short. First check that the free device wires at boxes are not inadvertently touching each other or the metal box. This might save you some time in tracking down a nonexistent short.

If your wiring is exposed, a careful visual check of the circuit should turn up the problem. If you still can't find the problem, or if wiring has already been covered, proceed as follows: Undo the wire splices at the next-to-last box on the circuit to open the circuit there. Then retest the circuit at the distribution center. Continue this procedure if necessary, moving closer to the source each time until the circuit tests out open. You have now isolated the short between the box where the circuit checked out open and the previous box (or at one of those boxes).

Outdoor wiring

You learned the ins and outs of installing indoor wiring in the preceding chapter. But chances are you'll also become involved with outdoor wiring, whether for security, for recreation, or to enhance your landscaping. Though principles are the same for both, some of the materials used in outdoor wiring are specially designed to resist the weather (and your garden hose). This chapter tells you what to look for.

Materials make the difference

Electrically, there is no difference between wiring inside and outside. The materials you use, however, are much different. Because outdoor wiring must survive the elements, materials are stronger and more corrosion resistant. Also, everything must fit exactly so gaskets will seal to prevent water from entering.

Boxes

Outdoor fixtures come in two types: so-called "driptight" fixtures that seal against vertically falling water and "watertight" ones that seal against water coming from any direction. It is important to know which type you need and which type you have purchased.

Driptight. Usually made of sheet metal and then painted, driptight fixtures often have shrouds or shields that deflect rain falling on top. A typical driptight subpanel is shown in drawing **82-A**. This unit is not waterproof and must be mounted where floods, or even "rain" from sprinklers below, cannot touch it.

Watertight. Fixtures designed to withstand *temporary* immersion or sprinkling are watertight. Made of cast aluminum, zinc-dipped iron, or bronze, these fixtures have threaded entries. All covers for watertight boxes are sealed with gaskets. Drawing **82-B** shows a watertight switch box.

Ground fault circuit interrupters

According to present electrical codes, any new outside receptacle (such as one used for a patio charcoal starter or for outdoor entertaining) must be protected with a GFCI (see pages 41 and 48). To

82-A: DRIPTIGHT SUBPANEL

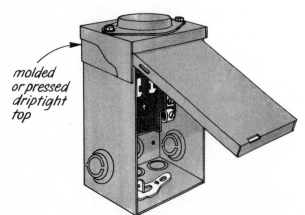

molded or pressed driptight top

82-B: WATERTIGHT SWITCH BOX

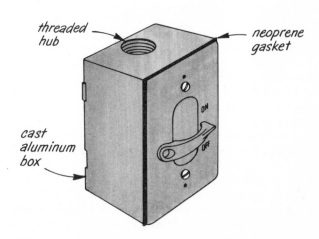

threaded hub

neoprene gasket

cast aluminum box

make your job easier if you are tapping into an existing circuit, you can buy a complete kit consisting of a cast aluminum box and cover and a GFCI receptacle.

Wiring materials

You can use type UF cable (see page 37) for direct burial to put in new receptacles and lights outdoors. The National Electrical Code permits burial of this cable only 12 inches deep for residential branch circuits if it is used for a 120 or 120/240-volt branch circuit. Do not make the mistake of using NM cable underground.

When burying UF cable, dig as deep as possible (12 inches minimum). Lay a redwood board on top of the cable before covering it with dirt (see drawing 83-A). This reduces the danger of spading through the cable at a later time.

83-A: BURYING OF CABLE

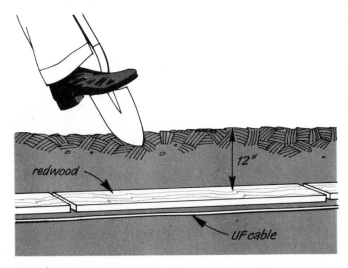

Conduit

In order of preference for the homeowner, conduits for underground use are rigid nonmetallic and rigid metal (see page 38). Rigid nonmetallic (PVC schedule 40) requires a separate grounding wire. It must be buried at least 18 inches deep unless it is covered with a concrete cap; in that case it can be buried at a lesser depth. One advantage of rigid nonmetallic conduit is that it does not corrode.

Rigid metal conduit will corrode and eventually disintegrate in certain types of soils. One advantage is that it may be directly buried only 6 or more inches deep. In addition, unless it's used to feed a swimming pool, it doesn't require a separate grounding wire.

Thinwall conduit (EMT) is not recommended to be used underground. Used with watertight couplings and connectors, however, EMT is useful in exposed locations above ground level. You must use some type of conduit to protect the conductors wherever they are exposed to physical abuse; EMT is a good choice in these situations.

Drawing 84-A illustrates the possible use of three types of conduit and UF cable. This should give you some ideas about how to plan your own underground installation.

Power sources: where and how

Electrically, extending an inside power source to the outside is the same as keeping all the wiring inside. You can tap into most switch, lighting, and receptacle outlet boxes as long as you don't tap into a switch or lighting outlet box that's at the end of a circuit run. Once again, the materials make the difference between indoor and outdoor wiring.

Possible power sources and ways to extend wiring from them are shown in drawing 83-B.

⚠ De-energize circuit before tapping into it.

83-B: OUTDOOR POWER SOURCES

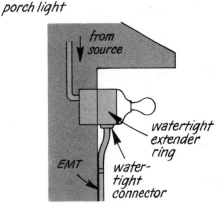

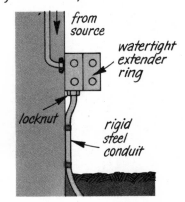

(Continued on next page)

box inside house

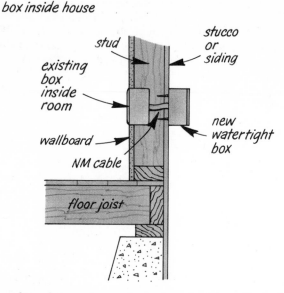

attic or crawl space

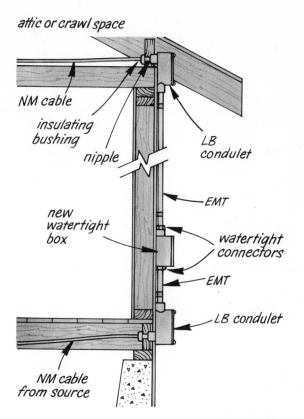

Remove faceplate and receptacle from inside wall. Drill ½-inch hole through back of plastic box, or punch out knockout. Drill hole through outside wall, using wood bit if siding, masonry bit if stucco. Screw watertight cast aluminum box to siding, or use screws and rawl plugs to secure box to stucco. Fish a short piece of NM cable between boxes and secure to boxes with cable clamps. Leave at least 8 inches of cable on each side. Make wire connections, then replace receptacle and faceplate.

84-A: TYPICAL OUTDOOR INSTALLATION

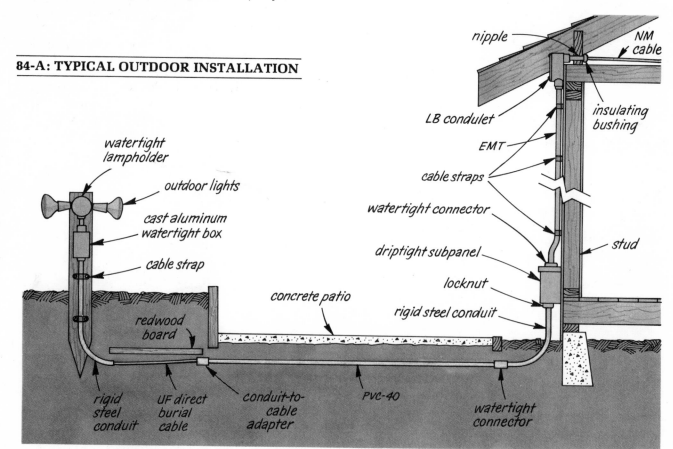

Cord-connected outdoor wiring

The foregoing material deals with installation of *permanent* outdoor wiring. However, you may be contemplating purchase of an above-ground pool or a hot tub. These are usually cord-connected.

Any of these pools should be used only on a GFCI-protected circuit, even if your home was built before GFCIs were required. In addition, if the pool or tub is not already equipped with a twist-lock plug, it is recommended that you change it. Then wire a special GFCI-protected twist-lock receptacle so the pool can be energized only by a GFCI-protected circuit. Drawing **85-A** shows a subpanel installation to feed a hot tub or above-ground pool.

85-A: TYPICAL HOT TUB INSTALLATION

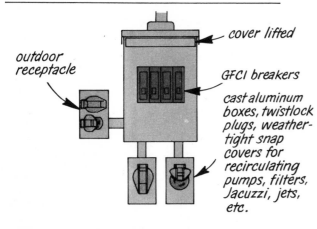

cover lifted

outdoor receptacle

GFCI breakers

cast aluminum boxes, twistlock plugs, weathertight snap covers for recirculating pumps, filters, Jacuzzi, jets, etc.

△ Always run a copper grounding conductor the same size as the circuit conductors with the circuit conductors from source to load, and for extra safety use GFCIs on all outdoor circuits.

Low-voltage lighting

Low-voltage lighting is a good choice for garden lighting. Operating at only 12 volts, low-voltage wiring is easier to install and doesn't present the dangers of 120 volts.

A transformer, usually driptight, is used to step the 120 volts down to 12 volts. This much reduced voltage is no longer as dangerous as 120 volts and does not require the special conduit and boxes of other outdoor wiring.

A watertight switch or receptacle is a good choice for a power source. Mount the transformer near the power source and then run direct burial cable from the low-voltage side of the transformer to the desired locations for your lights. The cable can be buried in the ground a few inches deep. However, to avoid accidentally spading through it, consider running the cable alongside structures, walks, and fences where you won't be likely to cultivate (see drawing **85-B**).

The low-voltage lighting fixtures attach directly to the wiring. Some fixtures simply clip onto the wire; others must be wired into the system. Low-voltage lights also come in a kit with a transformer of the proper size for the number of lights. Be sure to use the exact size of wire called for in the instructions.

85-B: TYPICAL LOW-VOLTAGE CIRCUIT

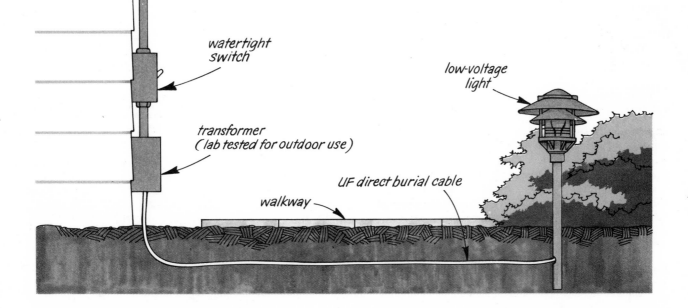

watertight switch

transformer (lab tested for outdoor use)

low-voltage light

UF direct burial cable

walkway

Electrical terms you should know

Alternating current. Current that regularly alternates its direction, flowing first in one direction and then in the opposite direction. Power companies generate alternating current to facilitate transmission of electricity over long distances. Abbreviated AC.

Ampacity. Coined word combining *ampere* and *capacity*. Expresses in amperes the current-carrying capacity of electric conductors.

Ampere. Unit used in measuring electrical current, based on the number of electrons flowing past a given point per second. Many elements of a wiring system are rated in amperes for the greatest amount of current they can safely carry. Abbreviated amp.

Bonding. Connecting metal components of an electrical system to form a continuous conductive path capable of handling any current likely to flow.

Branch circuit. Any one of many separate circuits distributing electricity throughout a house from the last overcurrent device protecting the circuit.

Circuit. Two or more wires providing a path for electric current to flow from source through some device using electricity (such as a light) and back to the source.

Circuit breaker. Safety switch installed in circuit to break electricity flow automatically when current exceeds a predetermined amount. Circuit breaker can be reset once "tripped."

Conductor. Trade name for electric wire.

Cycle. One complete reversal of alternating current: a forward flow (positive alternation) and backward flow (negative alternation). Ordinary household current has 60 cycles per second.

Fuse. Safety device installed in circuit to protect against an overload of current. Designed to "blow," (opening circuit and stopping flow of electricity) when current exceeds a predetermined amount. Fuses can't be reused once blown.

Ground. n. Any conducting body that serves in place of the earth, such as a metal cold-water pipe or a metal rod driven solidly into the earth. Sometimes called "grounding electrode." v.t. To connect any part of an electrical wiring system to the ground.

Grounding electrode conductor. Conductor connecting neutral bus bar of service entrance panel to ground. Sometimes called "ground wire."

Grounding wire. Conductor that grounds a metal component but does not carry current during normal operation. Returns current to source in order to open circuit breaker or fuse if metal component accidentally becomes electrically alive.

Hot bus bars. Solid metal bars connected to main power source in service entrance panel and subpanel. Branch circuit hot wires are connected to them.

Hot wire. Ungrounded conductor carrying electrical current. Usually identified by black or red insulation, but may be any color other than white, gray, or green.

Insulation. Sheathing of nonconducting material used to cover wires.

Kilowatt. Unit of electrical power equal to 1,000 watts. Abbreviated kw.

Kilowatt-hour. Unit used for metering and selling electricity. One kilowatt-hour equals 1,000 watts used for one hour (or any equivalent, such as 500 watts used for two hours). Abbreviated kwh.

Neutral bus bar. Solid metal bar in service entrance panel or subpanel which provides a common terminal for all neutral wires.

In service entrance panel, neutral bus bar is bonded to metal cabinet and is in direct contact with earth through grounding electrode conductor. All neutral and grounding wires are connected to this bus bar.

In subpanel, only neutral wires are connected to neutral bus bar, which "floats" in metal cabinet (it is not bonded).

Neutral wire. Grounded conductor that completes a circuit by providing a return path to the source. Except for a few switching situations, neutral wires must never be interrupted by fuse, circuit breaker, or switch. Always identified by white or gray insulation.

Ohm. The unit of measurement for electrical resistance or impedance.

Outlet. Box where wires terminate for connection to fixtures that consume electricity.

Overcurrent protection device. Fuse or circuit breaker that shuts off electricity flow when a conductor carries more than a